READY-TO-USE PHYSICAL SCIENCE ACTIVITIES FOR GRADES 5-12

Mark J. Handwerker, Ph.D.

THE CENTER FOR APPLIED
RESEARCH IN EDUCATION
West Nyack, New York 10994

Library of Congress Cataloging-in-Publication Data

Handwerker, Mark J.
Ready-to-use physical science activities for grades 5–12 / Mark J. Handwerker.
p. cm.—(Secondary science curriculum activities library)
ISBN 0-87628-437-3
1. Physical sciences—Study and teaching (Secondary) 2. Physical sciences—Study and teaching (Higher) I. Title. II. Series.
Q181.H218 1998
500.2′078—dc21 98-22200
CIP

Printed in the United States of America

10 9 8 7 6 5 4 3 2

ISBN 0-87628-437-3

THE CENTER FOR APPLIED RESEARCH IN EDUCATION
West Nyack, NY 10994
A Simon & Schuster Company

On the World Wide Web at http://www.phdirect.com

About This Resource

Ready-to-Use Physical Science Activities for Grades 5–12 is designed to help you teach basic science concepts to your students while building their appreciation and understanding of the work of generations of curious scientists. Although The Scientific Method remains the most successful strategy for acquiring and advancing the store of human knowledge, science is—for all its accomplishments—still merely a human endeavor. While the benefits of science are apparent in our everyday lives, its resulting technology could endanger the survival of the species if it is carelessly applied. It is therefore essential that our students be made aware of the nature of scientific inquiry with all its strengths and limitations.

A primary goal of every science instructor should be to make their students "science literate." After completing a course of study in any one of the many scientific disciplines, students should be able to:

1. appreciate the role played by observation and experimentation in establishing scientific theories and laws,
2. understand cause-and-effect relationships,
3. base their opinions on fact and observable evidence—not superstitions or prejudice, and
4. be willing to change their opinion based on newly acquired evidence.

Scientific theories come and go as new observations are made. During the course of instruction, teachers should emphasize the "process" of science as well as the relevance of pertinent facts.

This volume of science activities was designed to accomplish all of the above, keeping in mind the everyday challenges faced by classroom instructors.

On Your Mark!

Begin by stimulating students' gray matter with basic scientific concepts through brainstorming and open discussion.

Get Set!

Kindle interest by making concepts real through demonstration and/or descriptive analogy.

Go!

Cement concepts into concrete form with exciting hands-on experience.

Each of the 15 teaching units in this volume of *Ready-to-Use Physical Science Activities for Grades 5–12* contains *four* 40–50 minute lessons and follows the same instructional sequence so that your students will always know what is expected of them. Each unit comes complete with the following:

- a **Teacher's Classwork Agenda for the Week** and **Content Notes for Lecture and Discussion,**
- a student **Fact Sheet** with **Homework Directions** on the back,
- four 40–50 minute **Lesson Plans,** each followed by its own **Journal Sheet** to facilitate student notetaking, and
- an end-of-the-unit **Review Quiz.**

Each unit has been tested for success in the classroom and is ready for use with minimal preparation on your part. Simply make as many copies of the Fact Sheet with Homework Directions, Journal Sheets, and Review Quizzes as you need for your class. And complete answer keys for the homework assignments and unit quiz are provided at the end of the Teacher's Classwork Agenda for the Unit.

Mark J. Handwerker

About the Author

Mark J. Handwerker (B.S., C.C.N.Y., Ph.D. in Biology, U.C.I.) has taught secondary school science for 15 years in the Los Angeles and Temecula Valley Unified School Districts. As a mentor and instructional support teacher, he has trained scores of new teachers in the "art" of teaching science. He is also the author/editor of articles in a number of scientific fields and the coauthor of an earth science textbook currently in use.

Dr. Handwerker teaches his students that the best way to learn basic scientific principles is to become familiar with the men and women who first conceived them. His classroom demonstrations are modeled on those used by the most innovative scientists of the past. He believes that a familiarity with the history of science, an understanding of the ideas and methods used by the world's most curious people, is the key to comprehending revolutions in modern technology and human thought.

Suggestions for Using These Science Teaching Units

The following are practical suggestions for using the 15 teaching units in this resource to maximize your students' performance.

Fact Sheet

At the start of each unit, give every student a copy of the **Fact Sheet** for that unit with the **Homework Directions** printed on the back. The Fact Sheet introduces content vocabulary and concepts relevant to the unit. You can check students' homework on a daily basis or require them to manage their own "homework time" by turning in all assignments at the end of the unit. Most of the homework assignments can be completed on a single sheet of standard-sized (8 ½" × 11") loose-leaf paper. Urge students to take pride in their accomplishments and do their most legible work at all times.

Journal Sheet

At the start of each lesson, give every student a copy of the appropriate **Journal Sheet** which they will use to record lecture notes, discussion highlights, and laboratory activity data. Make transparencies of Journal Sheets for use on an overhead projector. In this way, you can model neat, legible, notetaking skills.

Current Events

Since science does not take place in a vacuum (and also because it is required by most State Departments of Education), make **Current Events** a regular part of your program. Refer to the brief discussion on "Using Current Events to Integrate Science Instruction Across Content Areas" in the Appendix.

Review Quiz

Remind students to study their Fact and Journal Sheets to prepare for the end-of-the-unit **Review Quiz.** The Review Quiz is a 15-minute review and application of unit vocabulary and scientific principles.

Grading

After completing and collectively grading the end-of-the-unit Review Quiz in class, have students total their own points and give themselves a grade for that unit. For simplicity's sake, point values can be awarded as follows: a neatly completed set of Journal Sheets earns 40 points; a neatly completed Homework Assignment earns 20 points; a neatly completed Current Event earns 10 points; and, a perfect score on the Review Quiz earns 30 points. Students should record their scores and letter grades on their individual copies of the **Grade Roster** provided in the Appendix. Letter grades for each unit can be earned according to the following point totals: A ≥ 90, B ≥ 80, C ≥ 70, D ≥ 60, F < 60. On the reverse side of the Grade Roster, students will find instructions for calculating their "grade point average" or "GPA." If they keep track of their progress, they will never have to ask "How am I doing in this class?" They will know!

Unit Packets

At the end of every unit, have students staple their work into a neat "unit packet" that includes their Review Quiz, Homework, Journal Sheet, Current Event, and Fact Sheet. Collect and examine each student's packet, making comments as necessary. Check to see that students have awarded themselves the points and grades they have earned. You can enter individual grades into your record book or grading software before returning all packets to students the following week.

You will find that holding students accountable for compiling their own work at the end of each unit instills a sense of responsibility and accomplishment. Instruct students to show their packets and Grade Roster to their parents on a regular basis.

Fine Tuning

This volume of *Ready-to-Use Physical Science Activities for Grades 5–12* was created so that teachers would not have to "reinvent the wheel" every week to come up with lessons that work. Instructors are advised and encouraged to fine tune activities to their own personal teaching style in order to satisfy the needs of individual students. You are encouraged to supplement lessons with your district's adopted textbook and any relevant audiovisual materials and computer software. Use any and all facilities at your disposal to satisfy students' varied learning modalities (i.e., visual, auditory, kinesthetic, etc.).

CONTENTS

About This Resource iii

Suggestions for Using These Science Teaching Units v

PS1 USING THE SCIENTIFIC METHOD / 1

Teacher's Classwork Agenda and Content Notes

Classwork Agenda for the Week . . . Content Notes for Lecture and Discussion . . . Answers to the End-of-the-Week Review Quiz

Fact Sheet with Homework Directions

Lesson #1
Students will define the term "science" and identify some of the many different scientific fields.
Journal Sheet #1

Lesson #2
Students will classify objects according to physical appearance and function.
Journal Sheet #2

Lesson #3
Students will list and discuss the steps of the scientific method.
Journal Sheet #3

Lesson #4
Students will use the scientific method to solve a problem by experiment.
Journal Sheet #4

PS1 Review Quiz

PS2 STUDYING POPULATIONS: MEANS, MEDIANS, AND MODES / 15

Teacher's Classwork Agenda and Content Notes

Classwork Agenda for the Week . . . Content Notes for Lecture and Discussion . . . Answers to the End-of-the-Week Review Quiz

Fact Sheet with Homework Directions

Lesson #1
Students will define the term "statistics" and explain why statistics is useful to scientists.
Journal Sheet #1

Lesson #2
Students will find the mean, median, and mode of a random group of measurements.
Journal Sheet #2

Lesson #3
Students will test an hypothesis using survey data.
Journal Sheet #3

Lesson #4
Students will use statistics to discuss the similarities and differences between two populations.
Journal Sheet #4

PS2 Review Quiz

PS3 MEASURING IN METRIC / 29

Teacher's Classwork Agenda and Content Notes

Classwork Agenda for the Week . . . Content Notes for Lecture and Discussion . . . Answers to the End-of-the-Week Review Quiz

Fact Sheet with Homework Directions

Lesson #1
Students will create a "standard" for measuring length and measure the length of objects using their standard.
Journal Sheet #1

Lesson #2
Students will compare standard units of measure for length and mass in the English and Metric Systems.
Journal Sheet #2

Lesson #3
Students will measure the length of objects in Metric Units of Measure.
Journal Sheet #3

Lesson #4
Students will use a balance to "mass" objects of different size and weight.
Journal Sheet #4

PS3 Review Quiz

PS4 MICRO TO MACRO: INTRODUCTION TO SCIENTIFIC NOTATION / 43

Teacher's Classwork Agenda and Content Notes

Classwork Agenda for the Week . . . Content Notes for Lecture and Discussion . . . Answers to the End-of-the-Week Review Quiz

Fact Sheet with Homework Directions

Lesson #1
Students will explain the meaning of base 10 and write small and large numbers in scientific notation.
Journal Sheet #1

Lesson #2
Students will add and substract small and large numbers using scientific notation and introduce the concept of significant figures.
Journal Sheet #2

Lesson #3
Students will multiply and divide small and large numbers using scientific notation.
Journal Sheet #3

Lesson #4
Students will continue multiplying and dividing small and large numbers using scientific notation.
Journal Sheet #4

PS4 Review Quiz

PS5 MEASURING LENGTH, AREA, AND VOLUME / 57

Teacher's Classwork Agenda and Content Notes

Classwork Agenda for the Week . . . Content Notes for Lecture and Discussion . . . Answers to the End-of-the-Week Review Quiz

Fact Sheet with Homework Directions

Lesson #1
Students will draw a three-dimensional cube on a two-dimensional surface.
Journal Sheet #1

Lesson #2
Students will calculate area from measures of length.
Journal Sheet #2

Lesson #3
Students will calculate volume from measures of length.
Journal Sheet #3

Lesson #4
Students will measure the volume of oddly shaped objects by water displacement.
Journal Sheet #4

PS5 Review Quiz

PS6 MEASURING MASS, VOLUME, AND DENSITY / 71

Teacher's Classwork Agenda and Content Notes

Classwork Agenda for the Week . . . Content Notes for Lecture and Discussion . . . Answers to the End-of-the-Week Review Quiz

Fact Sheet with Homework Directions

Lesson #1
Students will demonstrate the difference between mass and weight and construct a simple cardboard balance.
Journal Sheet #1

Lesson #2
Students will measure the mass of a variety of "weightless" objects.
Journal Sheet #2

Lesson #3
Students will measure the volume of oddly shaped objects.
Journal Sheet #3

Lesson #4
Students will calculate the density of a variety of objects.
Journal Sheet #4

PS6 Review Quiz

PS7 SPEED, VELOCITY, AND ACCELERATION / 85

Teacher's Classwork Agenda and Content Notes

Classwork Agenda for the Week . . . Content Notes for Lecture and Discussion . . . Answers to the End-of-the-Week Review Quiz

Fact Sheet with Homework Directions

Lesson #1
Students will describe how speed is different than velocity.
Journal Sheet #1

Lesson #2
Students will calculate the velocity of moving objects.
Journal Sheet #2

Lesson #3
Students will graph the positions of objects accelerating and those moving at a constant velocity.
Journal Sheet #3

Lesson #4
Students will measure and calculate the acceleration of moving objects.
Journal Sheet #4

PS7 Review Quiz

PS8 SIR ISAAC NEWTON AND THE LAWS OF MOTION / 99

Teacher's Classwork Agenda and Content Notes

Classwork Agenda for the Week . . . Content Notes for Lecture and Discussion . . . Answers to the End-of-the-Week Review Quiz

Fact Sheet with Homework Directions

Lesson #1
Students will demonstrate that massive objects have inertia and show how the force of friction interferes with the motion of objects.
Journal Sheet #1

Lesson #2
Students will show the relationship between force, mass, and acceleration.
Journal Sheet #2

Lesson #3
Students will show how the force on an object, its mass, or acceleration can be calculated using Newton's Second Law of Motion.
Journal Sheet #3

Lesson #4
Students will explain why the Law of Conservation of Momentum makes rocket flight possible.
Journal Sheet #4

PS8 Review Quiz

PS9 GALILEO, NEWTON, AND THE LAW OF GRAVITY / 113

Teacher's Classwork Agenda and Content Notes

Classwork Agenda for the Week . . . Content Notes for Lecture and Discussion . . . Answers to the End-of-the-Week Review Quiz

Fact Sheet with Homework Directions

Lesson #1
Students will demonstrate how to find the center of gravity of an oddly shaped object.
Journal Sheet #1

Lesson #2
Students will show that objects accelerate at a constant rate toward the earth.
Journal Sheet #2

Lesson #3
Students will show that objects accelerate toward the surface of the earth at a constant rate regardless of their masses.
Journal Sheet #3

Lesson #4
Students will find the force of gravity between two massive objects using Sir Isaac Newton's formula for calculating gravitational force.
Journal Sheet #4

PS9 Review Quiz

PS10 ARCHIMEDES AND BERNOULLI: FLUID PRESSURE / 127

Teacher's Classwork Agenda and Content Notes

Classwork Agenda for the Week . . . Content Notes for Lecture and Discussion . . . Answers to the End-of-the-Week Review Quiz

Fact Sheet with Homework Directions

Lesson #1
Students will demonstrate how pressure is related to force.
Journal Sheet #1

Lesson #2
Students will demonstrate that atmospheric pressure can change the shape of objects.
Journal Sheet #2

Lesson #3
Students will demonstrate how to use Bernoulli's Principle to explain how flight is possible.
Journal Sheet #3

Lesson #4
Students will demonstrate how to use Archimedes' Principle to explain why some objects float while others sink.
Journal Sheet #4

PS10 Review Quiz

PS11 FORCE, WORK, AND SIMPLE MACHINES / 141

Teacher's Classwork Agenda and Content Notes

Classwork Agenda for the Week . . . Content Notes for Lecture and Discussion . . . Answers to the End-of-the-Week Review Quiz

Fact Sheet with Homework Directions

Lesson #1
Students will explain the scientific meaning of the term "work" and identify two basic kinds of machines.
Journal Sheet #1

Lesson #2
Students will demonstrate how a lever produces a mechanical advantage.
Journal Sheet #2

Lesson #3
Students will demonstrate how an inclined plane produces a mechanical advantage.
Journal Sheet #3

Lesson #4
Students will demonstrate how a pulley produces a mechanical advantage.
Journal Sheet #4

PS11 Review Quiz

PS12 MACHINE EFFICIENCY AND POWER / 155

Teacher's Classwork Agenda and Content Notes

Classwork Agenda for the Week . . . Content Notes for Lecture and Discussion . . . Answers to the End-of-the-Week Review Quiz

Fact Sheet with Homework Directions

Lesson #1
Students will explain the relationship between "work" and "power."
Journal Sheet #1

Lesson #2
Students will draw a pulley system with a mechanical advantage of 50 and construct it to demonstrate that no machine is ever 100% efficient.
Journal Sheet #2

Lesson #3
Students will show how a wheel and axle assembly produces a mechanical advantage and how gears work according to the same principle.
Journal Sheet #3

Lesson #4
Students will calculate the mechanical advantage of a bicycle and discuss why the bicycle is one of the most efficient machines.
Journal Sheet #4

PS12 Review Quiz

PS13 ENERGY / 169

Teacher's Classwork Agenda and Content Notes

Classwork Agenda for the Week . . . Content Notes for Lecture and Discussion . . . Answers to the End-of-the-Week Review Quiz

Fact Sheet with Homework Directions

Lesson #1
Students will set up a pendulum and discuss how potential and kinetic energy can be transformed from one to the other.
Journal Sheet #1

Lesson #2
Students will construct a model of a steam turbine.
Journal Sheet #2

Lesson #3
Students will explain the purpose of each stroke of the standard 4-stroke internal combustion engine used in most modern automobiles.
Journal Sheet #3

Lesson #4
Students will pick out pertinent information from an electric bill.
Journal Sheet #4

PS13 Review Quiz

PS14 ELECTROMAGNETISM / 183

Teacher's Classwork Agenda and Content Notes

Classwork Agenda for the Week . . . Content Notes for Lecture and Discussion . . . Answers to the End-of-the-Week Review Quiz

Fact Sheet with Homework Directions

Lesson #1
Students will trace the pattern created by a magnetic field and discuss the basic laws of magnetism.
Journal Sheet #1

Lesson #2
Students will determine the types of materials influenced by a magnetic field.
Journal Sheet #2

Lesson #3
Students will demonstrate that magnetism and electricity are two aspects of a single force.
Journal Sheet #3

Lesson #4
Students will build an electromagnet.
Journal Sheet #4

PS14 Review Quiz

PS15 STATIC AND CURRENT ELECTRICITY / 197

Teacher's Classwork Agenda and Content Notes

Classwork Agenda for the Week . . . Content Notes for Lecture and Discussion . . . Answers to the End-of-the-Week Review Quiz

Fact Sheet with Homework Directions

Lesson #1
Students will create a static electric field that can be detected with the use of an electroscope.
Journal Sheet #1

Lesson #2
Students will draw electrical circuit diagrams for a simple series circuit and explain the relationship between amperage, voltage, and resistance in a circuit.
Journal Sheet #2

Lesson #3
Students will construct a simple parallel circuit and discuss the advantages of this kind of circuit over a series circuit.
Journal Sheet #3

Lesson #4
Students will construct a simple electric motor.
Journal Sheet #4

PS15 Review Quiz

APPENDIX / 211

Extra Journal Sheet • 213

Grade Roster • 214

How to Calculate Your Grade Point Average • 215

Using Current Events to Integrate Science Instruction Across Content Areas • 216

PS1

USING THE SCIENTIFIC METHOD

Teacher's Classwork Agenda and Content Notes

Classwork Agenda for the Week

1. Students will define the term "science" and identify some of the many different scientific fields.
2. Students will classify objects according to physical appearance and function.
3. Students will list and discuss the steps of the scientific method.
4. Students will use the scientific method to solve a problem by experiment.

Content Notes for Lecture and Discussion

Ask your students the following question at the start of Lesson #1: "How long have people used science to gather knowledge about the world around them?" Their answers to this question will reflect their level of historical knowledge and prompt discussion on the nature and purpose of scientific investigation. Ask your students to define the term "science." Point out that science began in the earliest days of humankind when people observed and recognized temporal patterns in nature: i.e., periodicity in the movements of the sun and moon, regularity in the change of the seasons. Science was practiced by the Chinese in the second millenium B.C. In India, the Middle East, Ancient Egypt, Ancient Greece, and the Americas, humankind's study of the patterns and regularities of nature were recorded and bequeathed to future generations. This legacy of natural philosophy we call "science."

The advance of science has since gone hand in hand with technology. Technology is defined as the "science of the mechanical and industrial arts." Early man manufactured tools of wood and stone. The first nails, lathes, and saws were used by Mesopotamians and ancient Egyptians as early as 3500 B.C. Later, the Hittites of Anatolia (ancient Turkey) discovered the forge to purify iron ore. This made possible the production of more practical tools and more destructive weapons of war. Technology allows people to "invent" practical applications for their ideas based on their observations of the natural world. Discuss the terms "mechanical" and "industrial" with the class. Have students refer to a dictionary to affirm their ideas and record their definitions as well as other highlights of class discussion on their Journal Sheet. Ask students to discuss the differences between science and technology. Tell students to list the familiar scientific disciplines and subdisciplines beginning with those defined on their Fact Sheet.

At the start of Lesson #2, point out that the rise of Modern Science coincided with the European Renaissance of the sixteenth and seventeenth centuries. The invention of the refracting telescope by **H. Lippershey** in Holland in 1608 and the development of new ideas to explain new (and old) observations began a whole new era of scientific investigation. The observation of the motions of the planets led to two opposing theories of the cosmos: The **Ptolemaic** and **Copernican Theories of The Solar System**. The names **Copernicus**, **Galileo**, and **Newton** should be introduced during this lesson. During this era, science became firmly entrenched in the minds of educated people as "an organized body of knowledge gathered by observation and experimentation." The Scientific Revolution led by these great scientists occurred during this historical period and resulted in the diffusion of the Scientific Method. Have students practice their observational skills by categorizing objects in the classroom according to physical appearance and practical function. Make sure students clearly define their categories and that members of each group agree on the definitions used in their category description. Clarity of language is an important part of scientific investigation. Ask students why this must be so. Reinforce the idea that the categories they have selected are not properties of the objects themselves but creations of the human mind!

At the start of Lesson #3, perform a simple demonstration like the one mentioned in the student Fact Sheet. Mix some baking soda and vinegar and ask students to record on their Journal Sheet the materials you use and the exact procedure that you follow. Ask them to draw some conclusions about the white powder and liquid you used. Will all white powders behave the same way when mixed with this liquid? And so on. Discuss the difference between a **demonstration** and an **experiment**. Emphasize that an experiment is a carefully controlled comparison of observations. Explain that scientific investigation begins with curiosity. Scientists observe nature then try to explain how nature behaves. The job of observing, comparing, and explaining how nature behaves generates a "cycle of investigation" called the Scientific Method. Have them summarize paragraph #4 in their Fact Sheet on their Journal Sheet.

The Scientific Method

Step #1: **Observe** and ask a question. Use one or more of the five senses (sight, hearing, smell, taste and touch) to describe a situation and ask a question whose answer will satisfy our curiosity. State a **problem**.

Step #2: Gather **information** related to the problem from other students or a textbook. Do some research at the library if that is convenient.

Step #3: Formulate an **hypothesis** to predict the outcome of a comparison. For example, say, "Not all white powders will result in the production of bubbles when mixed with this liquid." An hypothesis is an "educated guess" about what a comparison will show.

Step #4: Design an **experiment** that eliminates at least one possible outcome of a comparison. In this case, find at least one powder that does not result in the production of bubbles when mixed with this liquid. Depending on the level of students, explain the meaning of dependent and independent variables in a controlled experiment. An **independent variable** is the object (or parameter) that is under the control of the experimenter (i.e., the choice of different powders). The **dependent variable** is the object (or parameter) that changes as a result of the experimental manipulation (i.e., the production of bubbles depends on which powder is mixed with the liquid). Have students design their experiment by completing Steps #1, 2, 3, 4a, and 4b on their Journal Work Sheet. In Lesson #4, they can perform the experiment and complete Step #4c.

Step #5: Draw a **conclusion** that explains the outcome of the experiment. The conclusion is a rejection or restatement of the hypothesis put forth in Step #3. Ask students if their conclusion generates more questions. For example: "Why does the first white powder, but not the second, produce bubbles when mixed with the liquid? Emphasize that **scientific theories** are the products of our imagination. They are designed to explain observations and provide us with a better understanding of the world around us. All theories undergo a constant evolution because more observations are being added to our store of knowledge all the time.

ANSWERS TO THE END-OF-THE-WEEK REVIEW QUIZ

1. physics
2. biology
3. true
4. true
5. comparison
6. B
7. A
8. E
9. C
10. D

Answers to the first 5-point question: sight, hearing, smell, taste, touch. Answers to the second 5-point activity will vary but should emphasize that a demonstration is merely an accomplished act, an observed event. An experiment involves a comparison of two or more observed events.

Name: ______________________ **Period:**________ **Date:** ____/____/____

PS1 FACT SHEET

USING THE SCIENTIFIC METHOD

CLASSWORK AGENDA FOR THE WEEK

(1) Define the term "science" and identify some of the many different scientific fields.
(2) Classify objects according to physical appearance and function.
(3) List and discuss the steps of the scientific method.
(4) Use the scientific method to solve a problem by experiment.

A **science** is an organized body of knowledge gathered by observation and experimentation. Scientists use their five senses—the senses of sight, hearing, smell, taste, and touch—to observe and examine the natural world. The tools of a scientist, like a microscope, telescope or space probe, help the scientist to "amplify" one or more of the five senses. After gathering information about the natural world a scientist arranges or classifies that information in an organized way. Organizing information helps the scientist to compare observations. An experiment is a carefully controlled comparison of observations.

There are many different scientific fields. **Physics**, **chemistry**, **biology**, and **geology** are the four most widely known fields of science. Physics is the study of the interactions between matter and energy. Chemistry is the study of how different forms of matter interact with one another. Biology is the study of life. And, geology is the study of the earth. All of these scientific disciplines have subdisciplines that help scientists to focus their attention on particular kinds of observations. **Nuclear physics** is a subdiscipline of physics. A nuclear physicist studies the interaction between the tiny particles that make up atoms. An atom is the basic unit of a chemical molecule. **Medicine** is a subdiscipline of biology. A medical doctor studies the things that affect the health of the human body.

Scientific investigation begins with curiosity. Scientists observe nature, then try to explain how nature behaves. The job of observing, comparing, and explaining how nature behaves generates a "cycle of investigation" called the **scientific method**.

The steps of the scientific method are as follows:

1. Observe and ask a question. State a **problem**.
2. Gather **information** related to the problem from others who may have already asked the same question. Do some research at the library.
3. Formulate an **hypothesis** to predict the outcome of a comparison. That is, make an educated guess about what your comparison will show. Base your guess on your library research.
4. Design an **experiment** that eliminates at least one possible outcome of your comparison.
5. Draw a **conclusion** that explains the outcome of your experiment.

Over the past several hundred years, the scientific method has become the most successful means of gathering knowledge. The explanations that scientists have come up with to explain their observations has led to the development of many different kinds of **technology**. Technology is the science of the mechanical and industrial arts.

Experimentation helps scientists to learn faster than mere demonstration or trial and error. A demonstration is not the same as an experiment. For example: mixing baking soda (a white powder) with vinegar (a mild acid) produces carbon dioxide (a clear gas) bubbles. But this type of activity is not an experiment. It is a demonstration! You could conclude from such a demonstration that all white powders produce bubbles when mixed with vinegar. However, to test your conclusion you would have to experiment. You would have to mix other white powders (like sugar or salt) with vinegar and compare the results of your efforts. Your comparison in this second activity would be an experiment.

Homework Directions

1. List 20 objects in your bedroom. Organize the objects into no less than three different categories according to their physical appearance (i.e., shape, color, texture). Organize the objects again into no less than three groups according to their function (i.e., things that protect my body from the weather, things that entertain, things that help me exercise, etc.).

   Assignment due: __________

2. Using the scientific method, find out which liquid floats best on water: syrup or oil. Be sure to list each step of the scientific method and detail what you did at each step.

   Example: *Step #1:* Problem: Which liquid floats best on water—syrup or oil?

   Step #2: Gather information: In the past I have observed that salad oil floats on vinegar which is made mostly of water.

   And so on . . .

   Assignment due: __________

3. Study for the end-of-the-week Review Quiz and be sure your Current Event is ready to turn in before you enter class.

______________________ Student's Signature

______________________ Parent's Signature

____/____/____ Date

PS1 Lesson #1

USING THE SCIENTIFIC METHOD

Work Date: ____/____/____

LESSON OBJECTIVE

Students will define the term "science" and identify some of the many different scientific fields.

Classroom Activities

On Your Mark!

Lead a discussion in which students decide how long people have used science to gather knowledge about the world around them. Point out the periodicities in nature (i.e., the movements of the sun and moon, the regularity in the change of seasons) that aroused the curiosities of ancient people. Discuss and define the term "technology" and ask students to discuss how science depends on technology.

Get Set!

Remind and show students how a dictionary can help to clarify word meanings.

Go!

Have students refer to a dictionary in order to affirm their own ideas and definitions. Have them record their definitions as well as other highlights of class discussion on Journal Sheet #1. Tell students to list and define familiar scientific disciplines and subdisciplines beginning with those defined on their Fact Sheet.

Materials

dictionary, textbook, and library facility if available

Name: ______________________________ Period:________ Date: ____/____/____

PS1 Journal Sheet #1

USING THE SCIENTIFIC METHOD

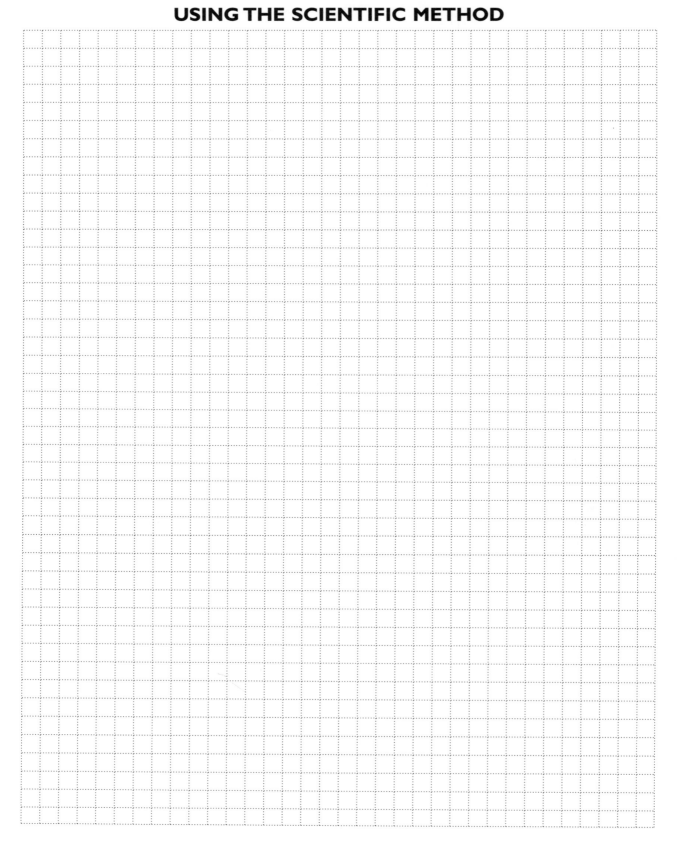

PS1 Lesson #2

USING THE SCIENTIFIC METHOD

Work Date: ____/____/____

LESSON OBJECTIVE

Students will classify objects according to physical appearance and function.

Classroom Activities

On Your Mark!

Lead a brief discussion of the senses we use to make observations (i.e., sight, hearing, smell, taste, and touch). Because people are dependent upon their limited senses they frequently observe events differently. In addition, people sometimes disagree about the causes of events even when they agree upon what they observe.

Get Set!

Have students examine the optical illusions on Journal Sheet #2 and discuss the answers to the questions posed with each illusion. In Figure A, the arrows are of equal length. Stare at Figures B, C, or D for a moment and the observer will "see" whatever they "wish" to see.

Go!

Have students practice their observational skills by categorizing objects in the classroom according to physical appearance and practical function. Make sure that students clearly define their categories since clarity of language is an important part of scientific investigation. Reinforce the idea that the categories they have selected are not properties of the objects themselves but convenient creations of their own minds.

Materials

classroom objects (i.e., desks, chairs, counters, wall posters, etc.), rulers

Name: ______________________ Period:________ Date: ___/___/___

PS1 Journal Sheet #2

USING THE SCIENTIFIC METHOD

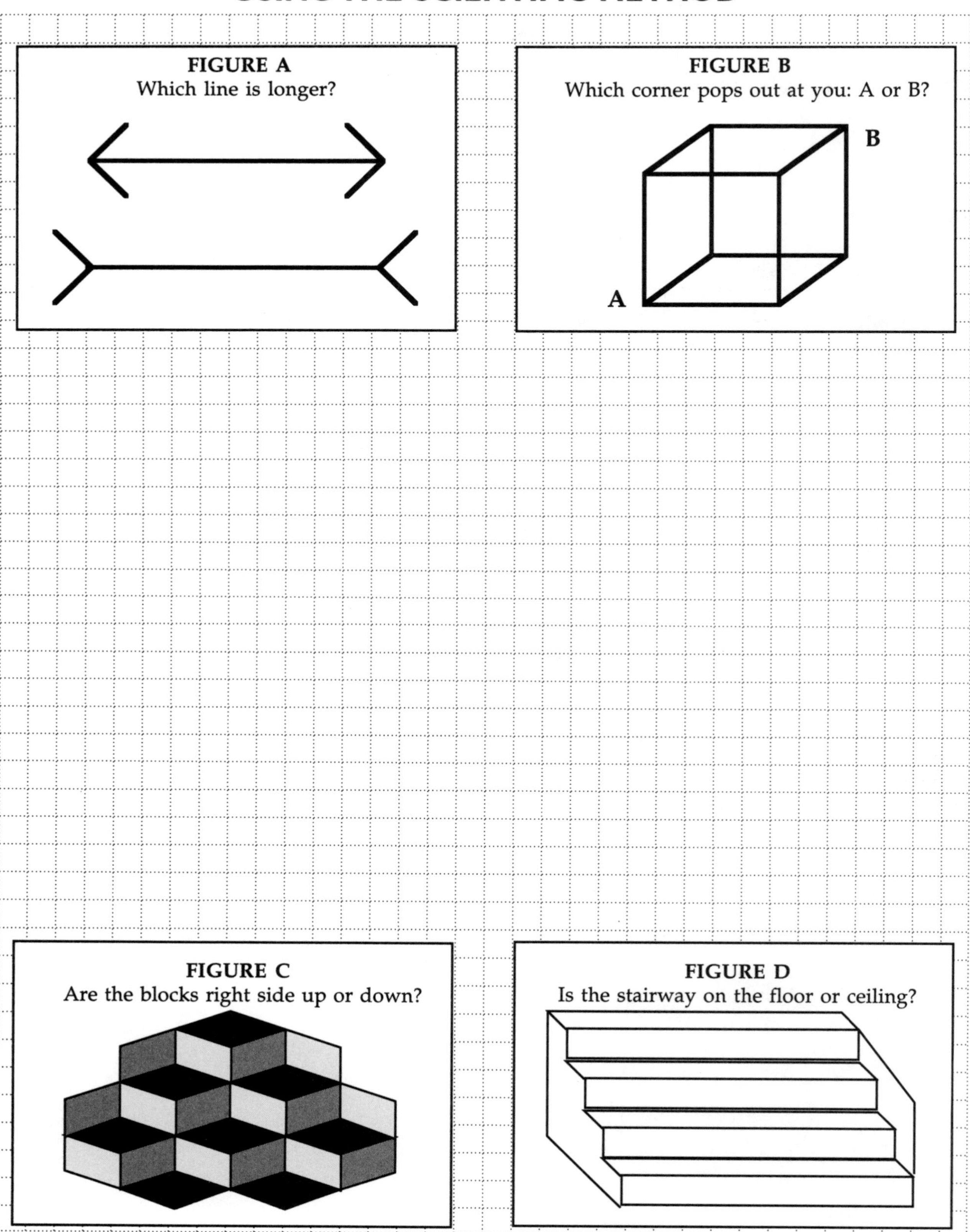

FIGURE A
Which line is longer?
FIGURE B
Which corner pops out at you: A or B?
B
A
FIGURE C
Are the blocks right side up or down?
FIGURE D
Is the stairway on the floor or ceiling?

USING THE SCIENTIFIC METHOD

Work Date: ____/____/____

LESSON OBJECTIVE

Students will list and discuss the steps of the scientific method.

Classroom Activities

On Your Mark!

Discuss the importance of making careful observations and why it is essential to report those observations as accurately as possible. After the demonstration, discuss the steps of the Scientific Method and why this method has been more successful than trial-and-error in increasing humankind's store of knowledge.

Get Set!

Ask students to record on Journal Work Sheet #3 the materials you use and the exact procedure you use in performing the following demonstration. First, pour a few tablespoons of vinegar into a small beaker. Second, mix a tablespoon of baking soda into the beaker. Have them draw conclusions about what they observed when the mixture overflowed with bubbles. Would changing the order in which the substances are mixed change the outcome of the demonstration? Try it. The answer—in this case—is no; but in some chemical reactions—such as the mixing of water and acid—the answer would be yes. Would all white powders behave the same as the baking soda when mixed with vinegar? If, yes—why? If, no—why not? Ask students to read their record of your procedure to one another. If they did a good job of observing and recording, they'd be able to duplicate exactly what you did using their description.

Go!

Have students summarize paragraph #4 in their Fact Sheet on their Journal Sheet #3. Using Journal Sheet #4, have students design an experiment that will answer the following question: "Do all white powders produce bubbles when mixed with vinegar? Have them complete steps #1, 2, 3, 4a, 4b, and 4c on Journal Sheet #4 in preparation for Lesson #4.

Materials

baking soda, vinegar, a small beaker

Name: ______________________________ Period:________ Date: ____/____/____

PS1 Journal Sheet #3

USING THE SCIENTIFIC METHOD

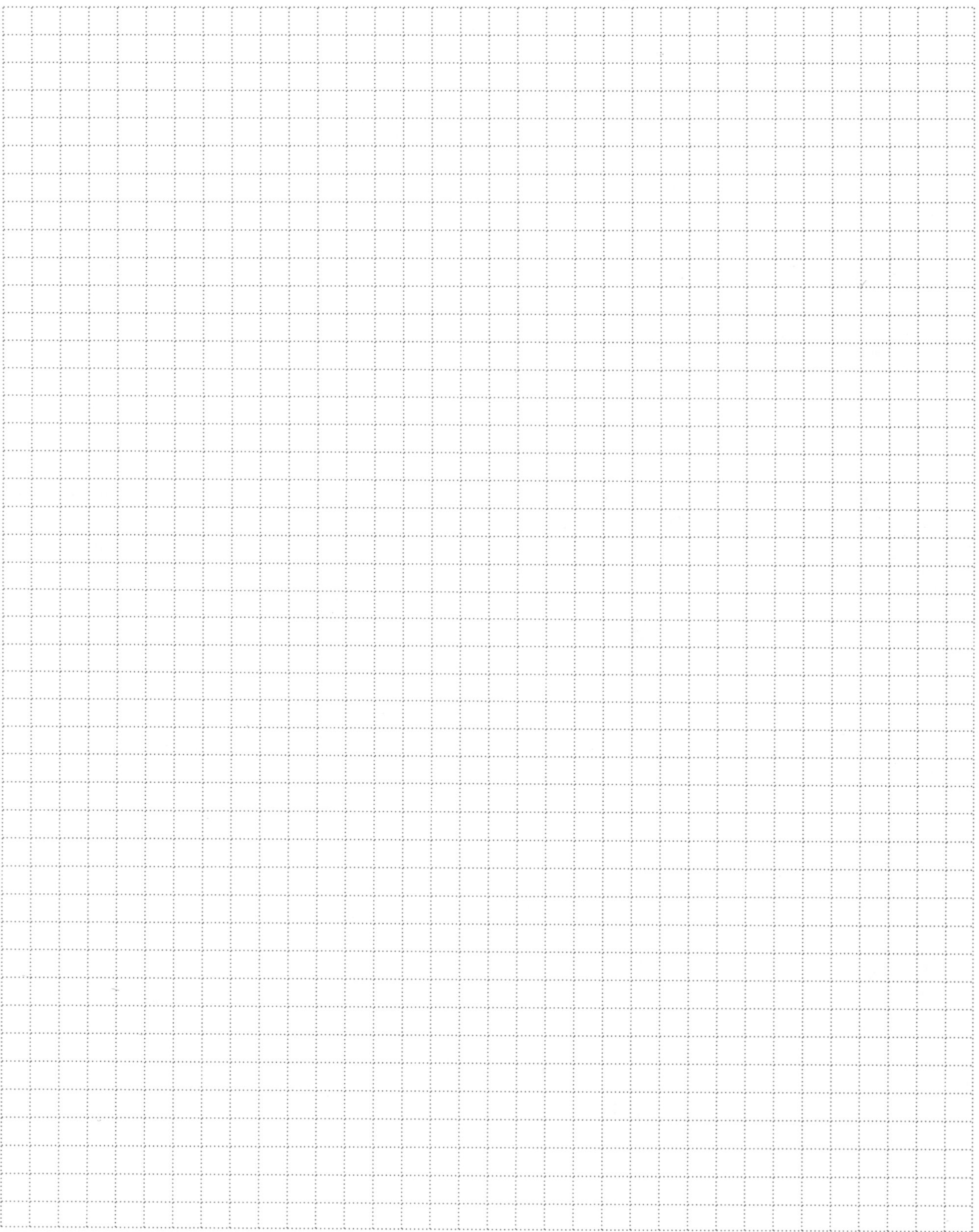

PS1 Lesson #4

USING THE SCIENTIFIC METHOD

Work Date: ____/____/____

LESSON OBJECTIVE

Students will use the scientific method to solve a problem by experiment.

Classroom Activities

On Your Mark!

Explain the importance of cleaning laboratory equipment and materials after every observation. Explain the importance of making careful measurements in preparing equal amounts of vinegar, baking soda, salt, and sugar used and how doing this helps to "control" their experiment.

Get Set!

Show students how to use a simple balance to measure the amounts of chemicals used.

Go!

Have students perform the experiment they designed in Lesson #3 and complete sections #4C and 5 on Journal Work Sheet #4.

Materials

balance, baking soda, salt, sugar, vinegar, small beakers, paper towel

Name: ______________________________ **Period:**_______ **Date:** ___/___/___

PS1 Journal Sheet #4

USING THE SCIENTIFIC METHOD

1. State the problem: ____

2. Gather information: ____

____

3. Form an hypothesis: ____

4. Experiment:
   a. list materials— ____

   ____

   b. explain step-by-step procedure— ______________________________

   ____

   ____

   c. describe what you observed during your experiment—________________

   ____

   ____

5. Form a conclusion: ____

____

Name: ______________________________ **Period:** ________ **Date:** ____/____/____

PS1 Review Quiz

Directions: Keep your eyes on your own work.
Read all directions and questions carefully.
THINK BEFORE YOU ANSWER!
Watch your spelling, be neat, and do the best you can.

CLASSWORK	(~40):	______
HOMEWORK	(~20):	______
CURRENT EVENT	(~10):	______
TEST	(~30):	______
TOTAL	(~100):	______

(A ≥ 90, B ≥ 80, C ≥ 70, D ≥ 60, F < 60)

LETTER GRADE: ______

TEACHER'S COMMENTS: ______________________________

____

USING THE SCIENTIFIC METHOD

TRUE–FALSE FILL-IN: If the statement is true, write the word TRUE. If the statement is false, change the underlined word to make the statement true. *10 points*

__________ 1. <u>Biology</u> is the study of the interactions between matter and energy.

__________ 2. <u>Physics</u> is the study of life.

__________ 3. <u>Chemistry</u> is the study of how materials interact.

__________ 4. <u>Geology</u> is the study of the earth.

__________ 5. An experiment always makes a(n) <u>mess</u>.

MATCHING: Put the letter of the step in the scientific method next to the activity done by Fred the Scientist. *10 points*

______ 6. Fred goes to the library to find out about ants and pill bugs.	(A) State a problem
______ 7. Fred asks: "Which group of insects is more sensitive to light: ants or pill bugs?"	(B) Gather information
______ 8. Based on his data, Fred decides that ants are more sensitive to light.	(C) Form a hypothesis
______ 9. Fred guesses that pill bugs are more sensitive to light than ants.	(D) Experiment
______ 10. Fred exposes ants and pill bugs to light and measures how fast it takes for each group to seek shelter in darkness.	(E) Draw a conclusion

PS1 Review Quiz *(cont'd)*

List the five senses that scientists use to make observations. *5 points*

__________ __________ __________ __________ __________

In a brief paragraph of no more than 30 words, explain the difference between a **demonstration** and an **experiment**. *5 points*

Student's Signature

Parent's Signature

___/___/___
Date

STUDYING POPULATIONS: MEANS, MEDIANS, AND MODES

TEACHER'S CLASSWORK AGENDA AND CONTENT NOTES

Classwork Agenda for the Week

1. Students will define the term "statistics" and explain why statistics is useful to scientists.
2. Students will find the mean, median, and mode of a random group of measurements.
3. Students will test an hypothesis using survey data.
4. Students will use statistics to discuss the similarities and differences between two populations.

Content Notes for Lecture and Discussion

Scientists do not have the luxury of studying every single object in the universe. **Statistics** provides a method of drawing valid inferences or conclusions about whole groups of individuals or objects by studying randomly sampled individuals or objects within a desired group. By gathering numerical data on the individuals within a group, the **statistician** can adequately, even reliably, describe the characteristics of an entire population.

At the start of Lesson #1, discuss with students the obvious difficulties in trying to study every individual within a population. Use the example of election returns mentioned in the Fact Sheet. Discuss the problems that might arise when using a limited, random sample to draw conclusions about a given population. For example, in a presidential election, taking a random sample from one of the east coast states may lead one to make the wrong prediction about the national election's outcome. In addition, it is difficult to contrast the characteristics of two populations of individuals when the individuals within each population are drastically different from one another. Statistical analysis takes these factors into account. Have students copy the definitions of the terms **mean**, **median**, and **mode** into their Journal Sheet. Explain how the drawings in the first work section of the Journal Sheet were obtained. The drawing shows two different populations of individuals. Each bar represents the number of individuals in a particular group. Each group scored differently on some arbitrary scale of measurement. For example, these graphs could be a comparision of two groups of students tested on a schoolwide exam. The bars may represent from left to right the grades of F, D, C, B, and A on the exam. The **normal distribution** may represent the whole student body at a particular grade level. The **skewed distribution** may represent the gifted students who always do well on tests. Explain the difference between a normal and a skewed distribution. Have students begin collecting data that will answer the following question: Which group of students (specify the age range of your class) is taller: boys or girls?

In Lessons #2 and #3, show students how to find the mean, median, and mode of a population. Use an example like the one assigned for homework. The principal descriptive value in statistics is the mean. It is the simple average of the accumulated scores. Show students how to find the median of a group of measures by performing the following demonstration. (1) Have students line up in rank order according to height from the shortest to the tallest. (2) Have them count off and remember their number. (3) Put the total number of students counted on the board and ask the class to identify the student(s) who occupy the middle of the line. In the case of an even number of students, two students will occupy the middle. (4) Explain that—excluding this middle person or persons—50% of the students in the line are shorter than this (these) individual(s) while 50% are taller. The number(s) assigned to the "middle" student(s) is the median of the population. Show students how to find the mode of the population. Point out that some students are exactly the same height (to within several centimeters). If student heights were measured to the nearest 10 centimeters, then more than several groups would likely be populated by more than one individual. The group with the most number of students is the mode of the population. In a normal distribution, the mean score, median score, and mode score of the population are the same value. Have students continue collecting data that will help them answer the question posed in Lesson #1. Compile and maintain data on the board from all classes of students between the ages 11 and 14.

At the start of Lesson #4, review the calculations involved in computing means, medians, and modes. Have students graph their information on Journal Sheet #4. Make sure they distinguish between boys and girls in their bar graph. To clearly see a difference between the populations (if indeed there is one) "overlap" the two distributions by alternating bars along the horizontal axis (i.e., boy bar—girl bar—boy bar—girl bar—etc.).

Statistical analysis is grounded in **probability theory**. And as such, statistics has become an important tool in the interpretation of physical phenomena. The **Theory of Quantum Mechanics,** for example, is based in large part on the understanding of statistics. According to quantum mechanical theory, the location of an electron orbiting an atomic nucleus, for example, can only be determined probabilistically. Statistics are more commonly used to describe the characteristics of populations and to test hypotheses like the one put forth in this unit (i.e., girls of a particular age range are taller than boys in the same age range). Hypothesis testing begins with inferring two (or more) possible outcomes of an experiment. Statistics is used to determine the effect (if any) of experimental manipulation on the dependent groups taking part in the experiment.

ANSWERS TO THE END-OF-THE-WEEK REVIEW QUIZ

1. B
2. A
3. B
4. C
5. E
6. dry
7. soaking wet
8. true
9. Section C
10. skewed

Name: ______________________________ Period:________ Date: ___/___/___

PS2 Fact Sheet

STUDYING POPULATIONS: MEANS, MEDIANS, AND MODES

CLASSWORK AGENDA FOR THE WEEK

(1) Define the term "statistics" and explain why statistics is useful to scientists.
(2) Find the mean, median, and mode of a random group of measurements.
(3) Test an hypothesis using survey data.
(4) Use statistics to discuss the similarities and differences between two populations.

All human beings share many of the same characteristics, yet no two are exactly alike. Even identical twins, although they look very similar, may have totally different personalities. If you look around your classroom you would have no trouble identifying some of the differences between you and your classmates. Human beings come in all shapes and sizes, skin colors, and physical abilities. You may also be struck by the similarities between yourself and your friends. The vast majority of human beings have two arms, two legs, two eyes, and one nose. No healthy human being has naturally green skin. Since it is not possible to study every individual in a population (i.e., the entire human race), scientists must always settle for studying a "sample" of that population. This fact is made obvious during election time. Scientists called **statisticians** work for the news department of your favorite television station. Statisticians are able to make very accurate predictions about election results. They base their predictions on "survey samples" taken from voters leaving the polls. *Collecting and analyzing information about individuals in a population in order to draw conclusions about the whole population* is called **statistics**. Statistics is the science of collecting and interpreting numerical data to draw conclusions about whole populations.

There are three important "statistical values" used in the analysis of scientific data: the mean, median, and mode. A **mean** is the average measure of a set of measures. A **median** is the midpoint (or middle number) in a series of measures. A **mode** is the most commonly occurring measure in a set of measures.

By comparing means, medians, and modes scientists can tell whether or not groups tested in an experiment are the same or different from one another. For example: What if you wanted to know which group of insects prefers shade to bright light: ants or spiders? It would be impossible to collect every ant and spider in your backyard, let alone the entire world. You would have to settle for a "sample population." Studying one ant and one spider would hardly be enough because no two ants are exactly alike just as individual human beings are not exactly alike. The best thing to do would be to collect a number of ants and spiders—perhaps a dozen or more. You could then use the Scientific Method and design an experiment to answer your question. You could measure the time it takes for individual ants and spiders to react to a sudden burst of light. Individual ants would probably react to the light at different speeds, but you could use math to find an "average" speed for all of the ants. You could then compare that average to the average speed obtained for your spiders.

The average measure of a set of measures is called the **mean**. To obtain the "mean of a population" simply add up the scores you measured for each individual, then divide by the number of individuals you sampled. The mean is the most reliable single measure used in statistics. However, the mean of a population may be misleading if the sample population is too small. For that reason, other statistical values are also compared.

The midpoint, or middle measure, in a group of individuals is called the **median**. To obtain the median measure of a population do the following: rank order the individuals from the lowest score to the highest score then find the score that lies in the middle of the group (i.e., the 50th percentile). Half of the sampled individuals lie below the median and half of the sampled scores lie above it. Have you ever taken a test and been told your "percentile" score?

Another valuable statistical measure is the mode. The most commonly occurring score in a set of measures is the **mode**. To find the mode of a population simply put the individuals with the same scores into the same group. The group with the most individuals is the mode of the population.

Statisticians use **graphs** to visualize the distribution of scores within a population. Using their graphs and the statistical values described above, the scientist can reliably describe whole populations by studying small samples of individuals.

Homework Directions

1. A teacher divided several science classes into two groups. One group, called the "Buddies," was told to study for the end-of-the-week quiz for 15 minutes per night with friends. The individuals belonging to the other group, called the "Loners," were told to study alone for 15 minutes per night. On their quiz, each student wrote the name of their group next to the amount of points earned on their quiz. The data gathered from this survey is shown in TABLE I.

   Use standard-sized graph paper to make a bar graph showing two distributions of quiz scores. Use red pencil or crayon to color the bars making up the Buddies' distribution of bars. Use blue pencil or crayon to color the bars making up the Loners' distribution of bars. Find the mean, median, and mode for each population. Based on your graph and statistics answer these two questions in a brief paragraph: Do the two populations appear the same or different? What can you conclude from this comparison?

TABLE I: THE BUDDIES' QUIZ SCORES – 20, 17, 21, 17, 18, 19, 18, 19, 19, 17, 19, 20, 10, 18, 20, 21, 21, 19, 18

THE LONERS' QUIZ SCORES – 14, 16, 14, 15, 16, 15, 15, 16, 17, 17, 15, 17, 14, 17, 16, 18, 18, 16, 18

Assignment due: __________

2. Study for the end-of-the-week Review Quiz and be sure your Current Event is ready to turn in before you enter class.

______________________________ Student's Signature

______________________________ Parent's Signature

____/____/____ Date

PS2 Lesson #1

STUDYING POPULATIONS: MEANS, MEDIANS, AND MODES

Work Date: ____/____/____

LESSON OBJECTIVE

Students will define the term "statistics" and explain why statistics is useful to scientists

Classroom Activities

On Your Mark!

Discuss and have students copy the definitions of the terms <u>mean</u>, <u>median</u>, and <u>mode</u> into their Journal Sheet #1. Be sure to explain how Figure A in Journal Sheet #1 was obtained (refer to Teacher's Classwork Agenda and Content Notes).

Get Set!

Demonstrate how to take measurements to the nearest centimeter using a metric ruler and point out that scientists around the world use the Metric System of Measurement exclusively over other methods of measurement. Review the steps of the Scientific Method.

Go!

Have students begin collecting data that will answer the following question: Which group of students (specify the age range of your class) is taller—boys or girls?

Materials

metric rulers, students

Name: ______________________________ Period: ________ Date: ____/____/____

PS2 Journal Sheet #1

STUDYING POPULATIONS

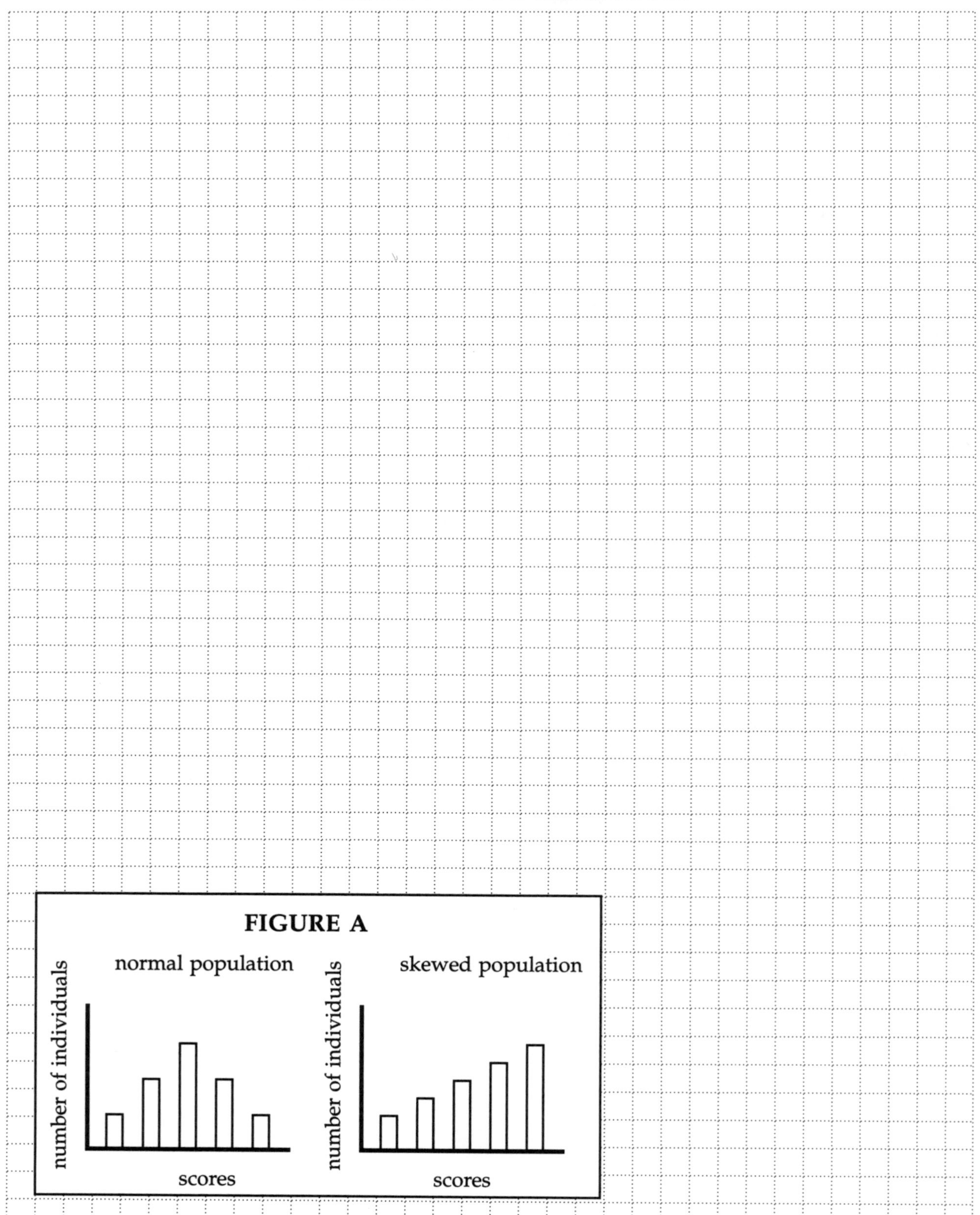

PS2 Lesson #2

STUDYING POPULATIONS: MEANS, MEDIANS, AND MODES

Work Date: ___/___/___

LESSON OBJECTIVE

Students will find the mean, median, and mode of a random group of measurements.

Classroom Activities

On Your Mark!

Reiterate the definitions of the terms mean, median, and mode as phrased in the student Fact Sheet.

Get Set!

Show students how to find the mean, median, and mode of a population. Use the following demonstration to help students grasp the meaning of the term "median." (1) Have students line up in rank order according to height from the shortest to the tallest. (2) Have them count off and remember their number. (3) Put the total number of students counted on the board and ask the class to identify the student(s) who occupy the middle of the line. In the case of an even number of students, two students will occupy the middle. (4) Explain that—excluding this middle person or persons—50% of the students in the line are shorter than this (these) individual(s) while 50% are taller. The number(s) assigned to the "middle" student(s) is the median of the population. Review the steps of the Scientific Method, again making sure that students have selected a hypothesis that will adequately help to answer the question posed in Lesson #1 (i.e., Are boys in our age range taller than girls the same age?). An hypothesis *states* a direct outcome! *It is not* a question. The purpose of the experiment is to test the statement.

Go!

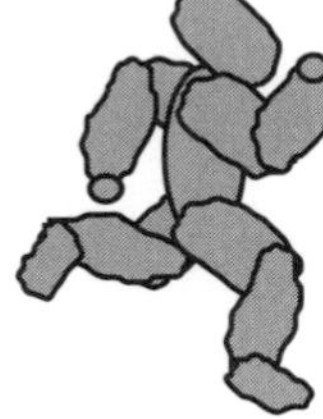

Have students continue collecting data to test their hypothesis.

Materials

metric rulers, students

Name: ______________________________ **Period:**________ **Date:** ____/____/____

PS2 Journal Sheet #2

STUDYING POPULATIONS

PS2 Lesson #3

STUDYING POPULATIONS: MEANS, MEDIANS, AND MODES

Work Date: ____/____/____

LESSON OBJECTIVE

Students will test an hypothesis using survey data.

Classroom Activities

On Your Mark!

Answer questions students have about the material covered in class so far. Answer questions students may have about the Homework Assignment.

Get Set!

Help students complete their data collection.

Go!

Have students continue collecting data to test the hypothesis posed by Lesson #1. When data collection is complete, assist students in graphing their results on Journal Sheet #4 and calculating the mean, median, and mode of the boy vs. girl populations.

Materials

metric rulers, students, calculators

Name: ______________________ Period:________ Date: ____/____/____

PS2 Journal Sheet #3

STUDYING POPULATIONS

STUDYING POPULATIONS: MEANS, MEDIANS, AND MODES

Work Date: ___/___/___

LESSON OBJECTIVE

Students will use statistics to discuss the similarities and differences between two populations.

Classroom Activities

On Your Mark!

Ask: "Why is a bar graph the most appropriate to use in this situation?" Review and discuss the difference between a bar graph and a line graph.

Get Set!

Give several examples of bar graphs and line graphs on the board.

Go!

Have students complete their graphs and the statistical analysis of their data on Journal Sheet #4. Lead a discussion to see how many students accepted/rejected their hypotheses. Point out that an hypothesis need not always be accepted. Whether it is accepted or rejected is not important. Either way, scientists learn from their experiment.

Materials

metric rulers, students, calculators

Name: ______________________________ Period:________ Date: ____/____/____

PS2 Journal Sheet #4

STUDYING POPULATIONS

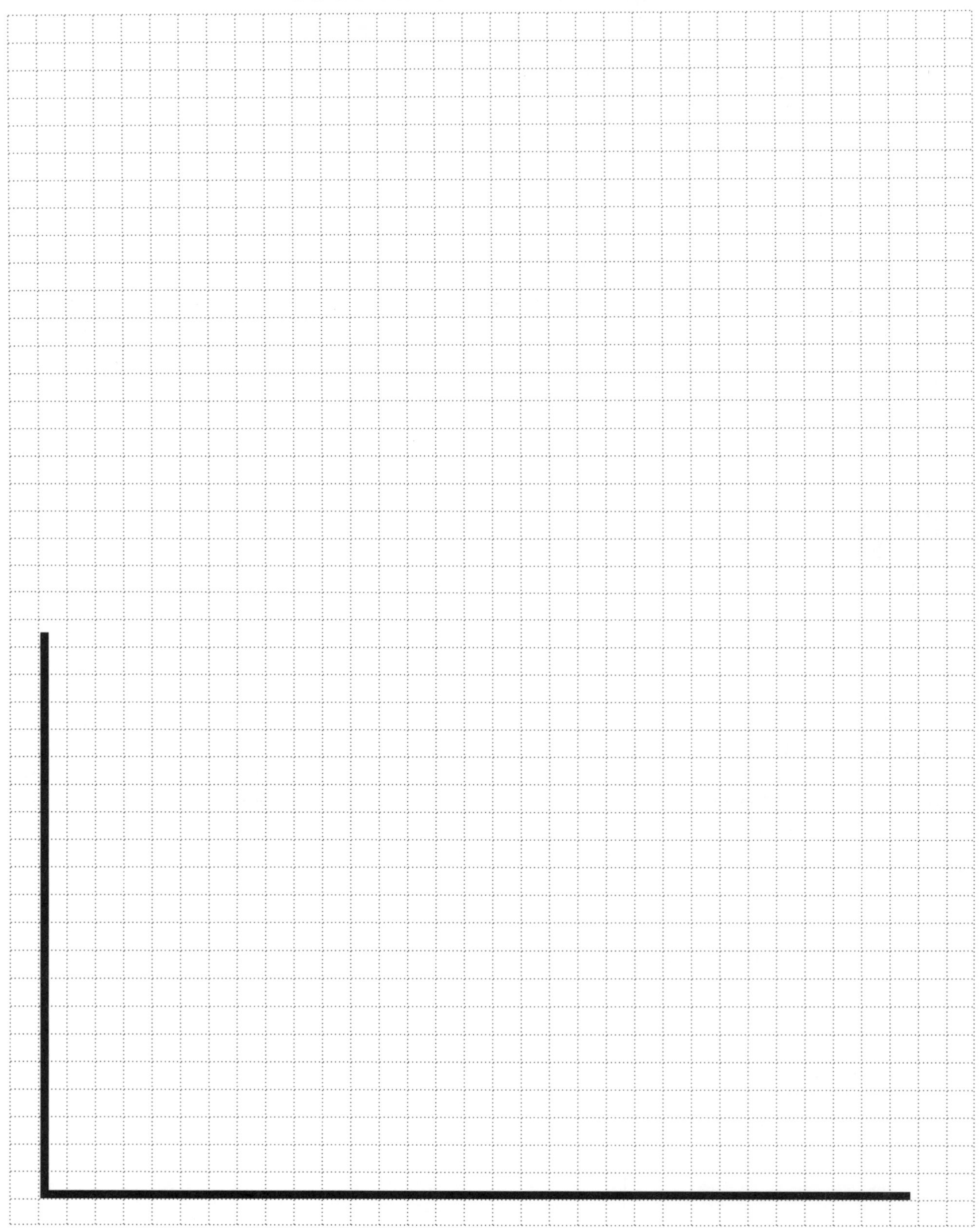

Name: ______________________ **Period:**________ **Date:** ___/___/___

PS2 Review Quiz

Directions: Keep your eyes on your own work.
Read all directions and questions carefully.
THINK BEFORE YOU ANSWER!
Watch your spelling, be neat, and do the best you can.

CLASSWORK	(~40):	______
HOMEWORK	(~20):	______
CURRENT EVENT	(~10):	______
TEST	(~30):	______
TOTAL	(~100):	______

(A ≥ 90, B ≥ 80, C ≥ 70, D ≥ 60, F < 60)

LETTER GRADE: ______

TEACHER'S COMMENTS: ____

____

STUDYING POPULATIONS: MEANS, MEDIANS, AND MODES

MULTIPLE CHOICE: Choose the letter of the choice that best answers the question or completes the sentence. *10 points*

_______ 1. Which term best describes a science that deals with the collecting and interpreting of numerical data?

(A) technology
(B) statistics
(C) algebra
(D) numerology
(E) none of the above

_______ 2. Which term best describes the average of a set of measures?

(A) mean
(B) median
(C) mode
(D) information
(E) data

_______ 3. Which term best describes the middle number in a series of measures?

(A) mean
(B) median
(C) mode
(D) information
(E) data

_______ 4. Which term best describes the most commonly occurring measure in a set of measures?

(A) mean
(B) median
(C) mode
(D) information
(E) data

_______ 5. How should the test scores in a normal population of individuals be distributed?

(A) fewer As than Bs
(B) fewer Bs than Cs
(C) more Cs than Ds
(D) more Ds than Fs
(E) in all the ways listed above

PS2 Review Quiz *(cont'd)*

PROBLEM

Read the following paragraph, refer to FIGURE I, and answer TRUE–FALSE FILL-IN questions #6 through #10. *20 points*

Alice collected pincher bugs and pill bugs to find out what type of soil each group of insects prefers. She divided a laboratory tray into three sections: Section A was covered in dry soil; Section B was covered in damp soil; Section C was covered with soaking wet soil. She gave each group of bugs ten minutes to choose the section of soil they liked most. At the end of ten minutes, Alice counted the number of bugs in each tray section. She graphed the information in Figure I.

FIGURE I

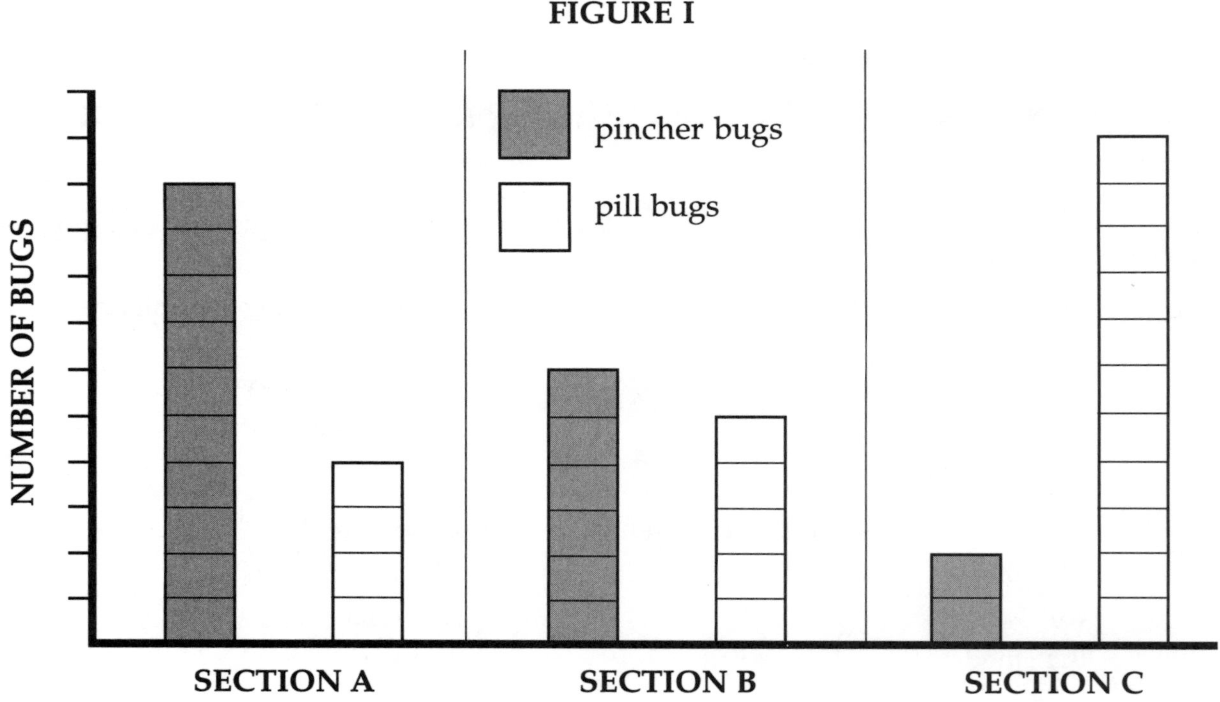

TRUE–FALSE FILL-IN: If the statement is true, write the word TRUE. If the statement is false, change the underlined word to make the statement true. *20 points*

_______________ 6. According to Figure I, pincher bugs prefer <u>damp</u> soil.

_______________ 7. According to Figure I, pill bugs prefer <u>dry</u> soil.

_______________ 8. The median score for the pincher bug population lies in <u>Section A</u>.

_______________ 9. The median score for the pill bug population lies in <u>Section B</u>.

_______________ 10. When it comes to choosing the type of soil where they'd like to live, pincher bugs are distributed in a <u>normal</u> population.

__________________________ Student's Signature

__________________________ Parent's Signature

____/____/____ Date

PS3

MEASURING IN METRIC

TEACHER'S CLASSWORK AGENDA AND CONTENT NOTES

Classwork Agenda for the Week

1. Students will create a "standard" for measuring length and measure the lengths of objects using their standard.
2. Students will compare standard units of measure for length and mass in the English and Metric Systems.
3. Students will measure the length of objects in Metric Units of Measure.
4. Students will use a balance to "mass" objects of different size and weight.

Content Notes for Lecture and Discussion

Scientific objectivity depends on **standard units of measure** that serve as a basis for comparing observations. Prompt students to discuss the problems that early humans must have had trying to describe the size of plants, animals, and other natural wonders to their "cavemates" without the help of a standard. Ask students to think of a variety of "standard units of measure" early humans might have used as a basis for comparison. The "cubit" is one such unit of measure defined as the length of a person's forearm from their elbow to fingertips. Conduct Lesson #1 to demonstrate the problems early observers encountered when comparing measured objects. Allow students to choose their own measurement standard: a body part, a worn down pencil, a paper strip of arbitrary length, etc. Conclude the activity by emphasizing the important function served by standard units of measure in all fields of science and technology. Instruct students to list the units of measure with which they are already familiar.

The United States, Canada, and Great Britiain are the only three nations of the world that do not officially use the **Metric System.** Review the relationship between common units of measure in the English System: the **foot**, **yard**, and **mile** for measures of **length**; the **ounce** and **pound** for measures of **weight**; the **pint**, **quart**, and **gallon** for measures of **liquid volume**. Explain that units of **time** are also based upon agreed standards of comparison. In the eighteenth century, mathematicians defined the **second** as 1/86,400 part of a mean solar day: the period from noon to noon on successive days. Today, a second is 9,192,631,770 vibrations of a cesium atom inside an atomic clock at the National Bureau of Standards in Boulder, Colorado. Conduct Lesson #2 to demonstrate the difficulty in converting English units of measure from one to another. This activity will also introduce students to standard units of measure in the Metric System. Use Charts #1 and #2 on the back of this week's student Fact Sheet to demonstrate the logic of the Metric System. Discuss how basing conversions on the number 10 makes it easy to change metric units of measure from one to another.

Lesson #3 gives students the opportunity to measure objects using metric units of measure. Continue to stress the importance of standards! Give students time to practice converting metric units of length from one to another. They will practice this further as part of this week's homework assignment.

Lesson #4 introduces students to the concept of **mass** which is defined as the amount of matter in an object. Demonstrate the use of a balance. Explain how early merchants used "standard units," such as carefully shaven stones or pieces of accurately molded metals and plaster, as objects to which amounts of grain and precious metals could be compared. Discuss some of the problems that might have been encountered in ancient marketplaces. Prompt a discussion that leads students to arrive at a clear distinction between the idea of mass and weight. **Weight** is defined as the force of gravity on an object. Ask students to think of situations in which an object

PS3 Content Notes *(cont'd)*

can be weightless yet retain its mass (i.e., in outer space, while falling, etc.). Explain that the basic unit of measure for mass in the Metric System is the **gram**. Define the gram as the amount of pure water that would fill a cubic centimeter volume at 4 degrees Celsius at sea level. Ask students to consider why eighteenth century mathematicians included measures of temperature and elevation in their definition of the gram. Conclude the activity by allowing students to convert metric units of mass from one to another using Fact Sheet Charts #1 and #2.

ANSWERS TO THE END-OF-THE-WEEK REVIEW QUIZ

1. E
2. E
3. C
4. E
5. E
6. length, distance
7. mass
8. foot
9. true
10. true
11. 86
12. 70
13. 4.8
14. 0.036
15. 0.000024

Name: ______________________________ Period:________ Date: ____/____/____

PS3 Fact Sheet

MEASURING IN METRIC

CLASSWORK AGENDA FOR THE WEEK

(1) Create a "standard" for measuring length and measure the lengths of objects using your standard.
(2) Compare standard units of measure for length and mass in the English and Metric Systems.
(3) Measure the length of objects in Metric Units of Measure.
(4) Use a balance to "mass" objects of different sizes and weights.

Imagine visiting the giant sequoia trees of Sequoia National Park, California, then describing that majestic sight to a friend. You might say something like, "The giant sequoias are the tallest trees I have ever seen." This description does not give your friend any exact information. It only describes a particular "quality" of the trees you observed, namely their height. It is a **qualitative observation.** Your friend might ask in reply: "How tall were they?" If you were not able to compare the size of the trees to some object familiar to your friend, you would have a tough time answering that quesion. You might answer: "The tallest sequoia I saw was as tall as a ten-story building." By comparing the height of the tree to something your friend can visualize (like a ten-story building) you make it easier for him or her to appreciate your description. In addition, your answer included a number to make your description more exact. You included a "quantity." This kind of observation is called a **quantitative observation**. In science, measuring the length, width, and height of objects involves the same kind of comparison. Scientists use a **standard unit of measure** to describe the dimensions of objects. A **standard** is anything used as a basis for comparison.

The traditional **English System of Measurement** uses the **inch**, **foot**, **yard**, and **mile** as units of measure for length. The problem with these units of measure is that they are not related to one another in any logical way. There are 12 inches in a foot; 3 feet in a yard; 5,280 feet or 1,760 yards in a mile. Measures of weight in the English System also have no logical relationship. There are 16 ounces in a pound and 2,000 pounds in a ton. Changing measures of length in miles to measures of length in feet; or measures of weight in ounces to measures of weight in tons, involves cumbersome math calculations.

In the eighteenth century, French scientists agreed to create a system of measurement that everyone would find easy to use. They calculated the distance from the earth's equator to the North Pole and divided that distance into 10 million parts. The size of the unit of measure they came up with is called a **meter**: the basic unit measure of the **Metric System**.

Other units of measure in the Metric System are compared to this "basic" standard. In the Metric System, the basic unit of measure for mass is the gram. **Mass** is the amount of matter in an object. By definition, one **gram** of any substance is equal to the amount of matter in a tiny cube of water (at sea level and 4 degrees Celsius) measuring one hundredth of a meter in length, width, and height. All units of measure in the Metric System can be compared to one another using simple calculations based on the number 10. In the two charts below you can see how easily metric units of measure are changed from one to another.

PS3 Fact Sheet *(cont'd)*

CHART #1: Prefixes used to relate metric units of measure.

Prefix	abbreviation	*compared to a meter (m)	
milli-	mm	one thousandth	(0.001)
centi-	cm	one hundredth	(0.01)
deci-	dm	one tenth	(0.1)
deka-	dam	ten	(10)
hecto-	hm	one hundred	(100)
kilo-	km	one thousand	(1,000)

*The same prefixes are used for measures of mass. For example, one thousandth of a gram is called one "milligram."

CHART #2: Comparing Metric Units of Measure using decimals.

	mm	cm	dm	m	dam	hm	km
mm =	1	0.1	0.01	0.001	0.0001	0.00001	0.000001
cm =	10	1	0.1	0.01	0.001	0.0001	0.00001
dm =	100	10	1	0.1	0.01	0.001	0.0001
m =	1,000	100	10	1	0.1	0.01	0.001
dam =	10,000	1,000	100	10	1	0.1	0.01
hm =	100,000	10,000	1,000	100	10	1	0.1
km =	1,000,000	100,000	10,000	1,000	100	10	1

Example shown in box: "One centimeter equals one hundredth of a meter."

Homework Directions

1. Using an English and a Metric ruler, measure the height of ten pieces of furniture or major appliances in your home in both inches and centimeters.
2. Change your measures from inches to yards and centimeters to meters.

Assignment due: __________

______________________ Student's Signature

______________________ Parent's Signature

____/____/____ Date

MEASURING IN METRIC

Work Date: ____/____/____

LESSON OBJECTIVE

Students will create a "standard" for measuring length and measure the lengths of objects using their standard.

Classroom Activities

On Your Mark!

Prompt students to discuss the problems early humans must have had trying to describe the size of plants, animals, and other natural wonders to their "cave-mates" without the help of a standard. Explain how scientific "objectivity" depends on the use of standard units of measure.

Get Set!

Introduce an ancient "standard unit of measure" called the the <u>cubit</u>: the length of a person's forearm from their elbow to fingertips. Ask students to discuss the difficulties they would have trying to build a house together using their individual cubits.

Go!

Allow students to choose their own measurement standard: a body part, a worn down pencil, a paper strip of arbitrary length, etc. Instruct them to measure the length of a variety of objects around the classroom using their "personal standard." Make sure they record their observations on their Journal Sheet #1. Instruct them to compare their units of measure with those of another group of classmates who used a different standard. Ask them to spend a few moments attempting to "convert" their unit of measure (i.e., the length of Mary's index finger) into the other group's unit of measure (i.e., the length of John's stubby pencil). For example: How many "index fingers" are equal to the length of one "stubby pencil?" Mary's notebook may be 3 index fingers, or 6 stubby pencils, in length. So, one-half of an index finger equals one stubby pencil. Conclude the activity by emphasizing the important function served by universally accepted, standard units of measure in all fields of science and technology. Instruct students to list the units of measure in the English and Metric Systems with which they are already familiar.

Materials

English and metric rulers, common objects

Name: ______________________ Period:________ Date: ____/____/____

PS3 Journal Sheet #1

MEASURING IN METRIC

PS3 Lesson #2

MEASURING IN METRIC

Work Date: ___/___/___

LESSON OBJECTIVE

Students will compare standard units of measure for length and mass in the English and Metric Systems.

Classroom Activities

On Your Mark!

Review the common units of measure in the English System (i.e., inch, foot, yard, mile). Ask: "Is there a logical relationship between these units?"

Get Set!

Lead an examination of English and Metric rulers. Begin with a comparison of the English and the Metric rulers at the bottom of Journal Sheet #2. Point out that there are approximately 2.54 centimeters (≈ 25.4 millimeters) in 1 inch. There are approximately 30.48 centimeters in 1 foot.

Go!

Tell students to measure the length of a classroom object (i.e., the width of a lab table or the height of a chair) in inches. Instruct them to convert that length to feet, yards, and miles. They will probably have difficulty with this as the relationship between English units is not logically based and they'll need to do a lot of cumbersome division and multiplication. Lead an examination of Charts #1 and #2 on their Fact Sheet. Have them measure the same object in centimeters then convert that figure to millimeters, meters, and kilometers. To do this, they need to refer to Chart #2 and learn how to move the decimal in their original measurement. Make sure they record all of their observations on their Journal Sheet #2.

Materials

English and metric rulers, calculators, classroom objects

Name: ______________________________ **Period:**________ **Date:** ____/____/____

PS3 Journal Sheet #2

MEASURING IN METRIC

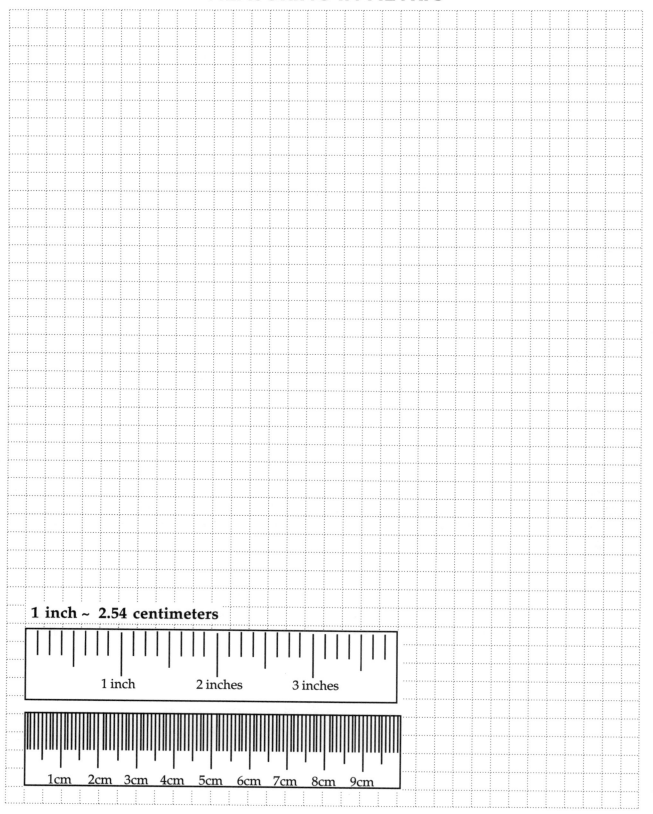

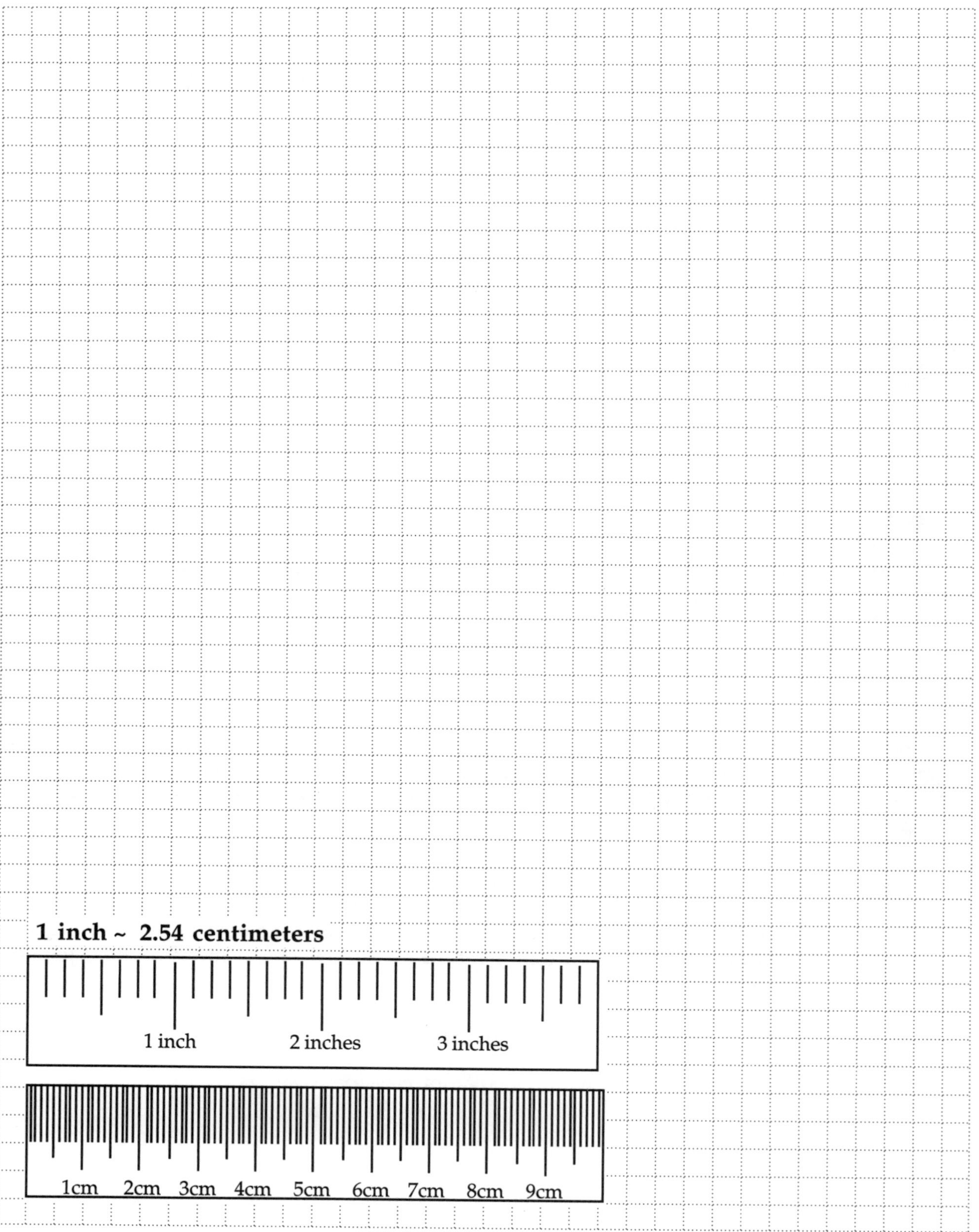

PS3 Lesson #3

MEASURING IN METRIC

Work Date: ____/____/____

LESSON OBJECTIVE

Students will measure the length of objects in Metric Units of Measure.

Classroom Activities

On Your Mark!

Review the difficulties encountered in converting English units of measure from one to another. Emphasize the advantages of using the Metric System, based on the number 10.

Get Set!

Do some simple conversions on the board and have students copy the examples on Journal Sheet #3.

Go!

Have students continue measuring the lengths of a variety of classroom objects. Allow them to practice converting metric units of measure from one to another using the charts on their Fact Sheet. At first, they may be confused about which way to move the decimal point, but with practice this skill will improve. Depending on the level of your students, you may decide not to hold them accountable for memorizing all of the metric prefixes. The most commonly used metric prefixes are milli-, centi-, and kilo-.

Materials

metric rulers, calculators, classroom objects

Name: ______________________ Period:________ Date: ___/___/___

PS3 Journal Sheet #3

MEASURING IN METRIC

PS3 Lesson #4

MEASURING IN METRIC

Work Date: ____/____/____

LESSON OBJECTIVE

Students will use a balance to "mass" objects of different size and weight.

Classroom Activities

On Your Mark!

Discuss the difference between mass and weight. Ask students to give examples of "weightless" objects that have mass (i.e., an astronaut in space, a person floating in a swimming pool, someone falling, etc.). Have students write the definition of a gram in their Journal Sheet #4.

Get Set!

Show students how ancient merchants "massed" objects like bags of salt or grain by comparing the weight of the salt or grain to "known" masses. Point out that the same metric prefixes used to relate small and large metric units of measure for length are used to compare metric units of mass. Demonstrate how to use a balance.

Go!

Have students "mass" a variety of small common objects. Use standard metric brass weights if available.

Materials

balance, common objects, standard metric brass weights

Name: ______________________ Period:________ Date: ____/____/____

PS3 Journal Sheet #4

MEASURING IN METRIC

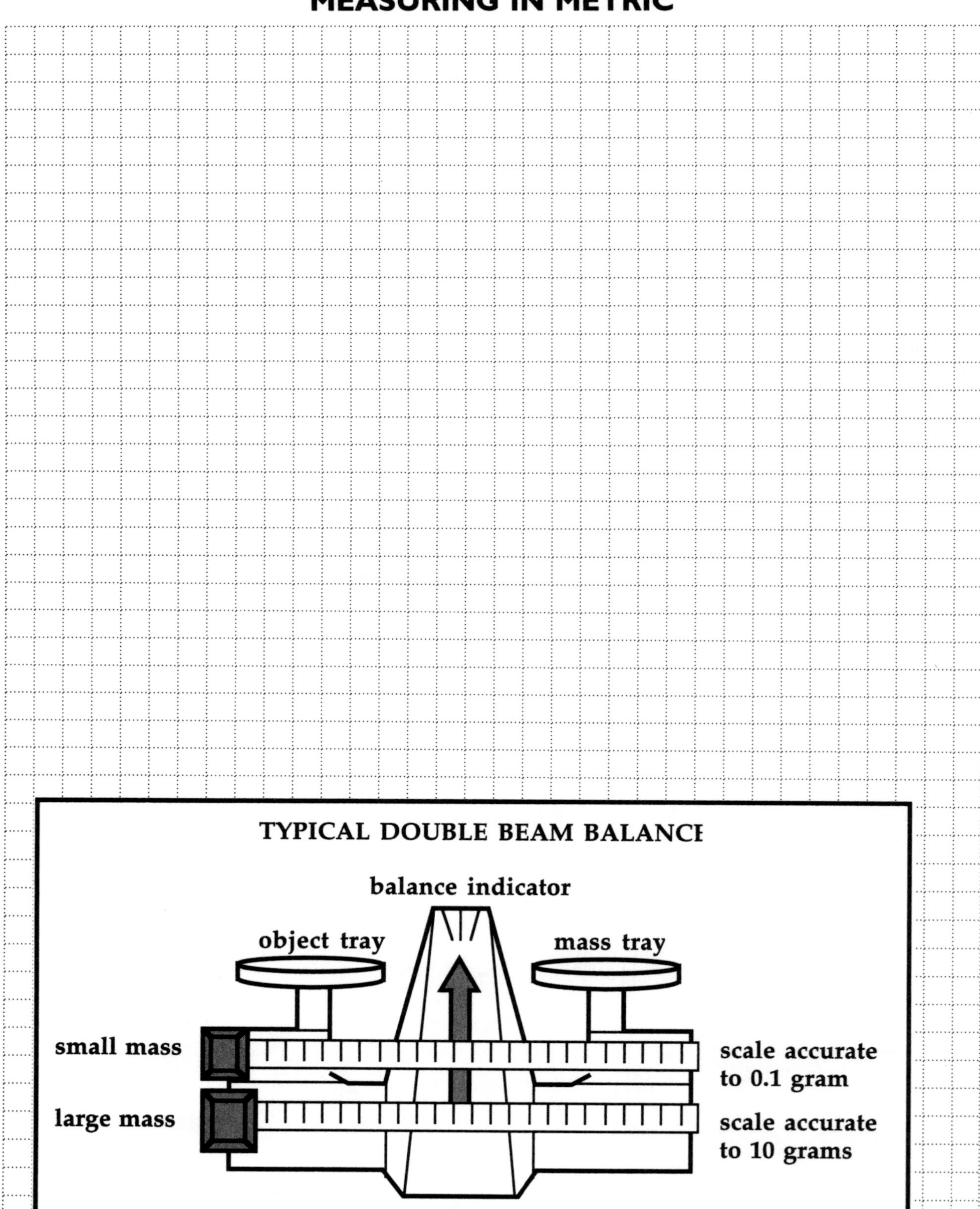

Name: ______________________ **Period:**________ **Date:** ___/___/___

PS3 Review Quiz

Directions: Keep your eyes on your own work.
Read all directions and questions carefully.
THINK BEFORE YOU ANSWER!
Watch your spelling, be neat, and do the best you can.

CLASSWORK	(~40):	_______
HOMEWORK	(~20):	_______
CURRENT EVENT	(~10):	_______
TEST	(~30):	_______
TOTAL	(~100):	_______

(A ≥ 90, B ≥ 80, C ≥ 70, D ≥ 60, F < 60)

LETTER GRADE: _______

TEACHER'S COMMENTS: ____

____

MEASURING IN METRIC

MULTIPLE CHOICE: In the space on the left, write the letter of the word or phrase that best completes the statement or answers the question. *10 points*

_____ 1. Which of the following IS NOT a metric unit of length?
(A) meter (C) yard (E) both B and C
(B) inch (D) both A and B

_____ 2. How many centimeters are there in 6 meters?
(A) 0.06 (C) 6.0 (E) 600
(B) 0.6 (D) 60

_____ 3. Which word best describes the meaning of a "standard?"
(A) majestic (C) comparison (E) mass
(B) easy (D) length

_____ 4. Which of the following is the SMALLEST unit of measure?
(A) mile (C) foot (E) millimeter
(B) kilometer (D) yard

_____ 5. Why did mathematicians invent the Metric System of measurement?
(A) to make measuring lengths easier
(B) to make comparing measures of length and mass easier
(C) to measure the distance from the earth's equator to the North Pole
(D) to confuse people who are not scientists
(E) none of the above are correct

PS3 Review Quiz *(cont'd)*

TRUE–FALSE FILL-IN: If the statement is true, write the word TRUE in the space on the left. If the statement is false, write a word or phrase in the space on the left that replaces the underlined word to make the statement true. *10 points*

________________ 6. A meter is a measure of mass.

________________ 7. A gram is a measure of length.

________________ 8. There are 12 inches in a yard.

________________ 9. There are 5,280 feet in one mile.

________________ 10. The milligram, centigram, and kilogram are units of measure in the Metric System.

PROBLEM

Fill in the blanks numbered 11 through 15 to indicate the correct measure shown on the metric ruler. *10 points*

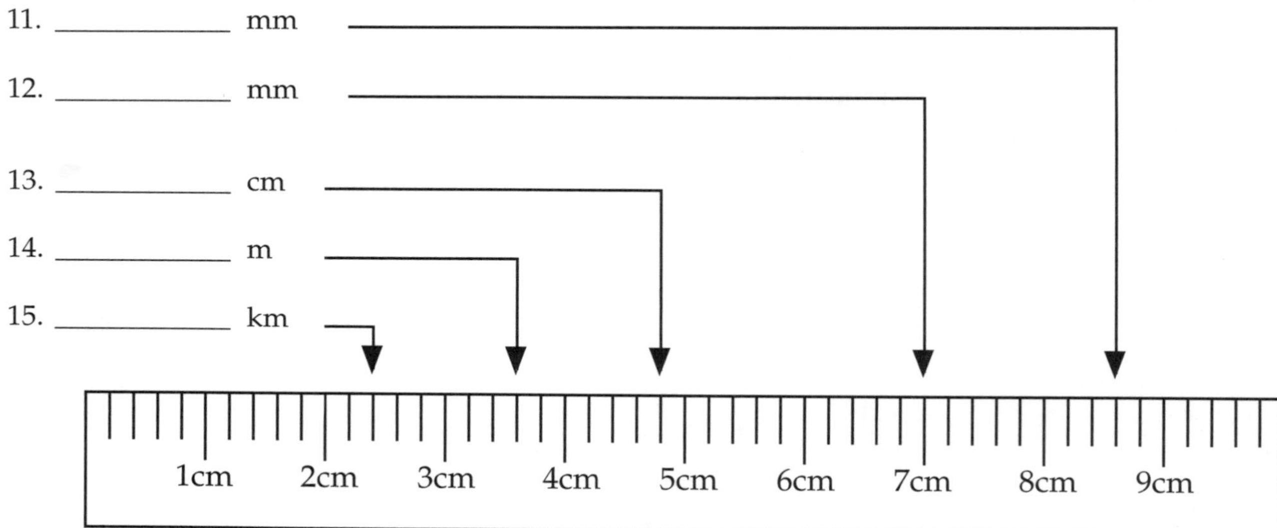

____________________ Student's Signature

____________________ Parent's Signature

____/____/____ Date

PS4 MICRO TO MACRO: INTRODUCTION TO SCIENTIFIC NOTATION

TEACHER'S CLASSWORK AGENDA AND CONTENT NOTES

Classwork Agenda for the Week

1. Students will explain the meaning of base 10 and write small and large numbers in scientific notation.
2. Students will add and subtract small and large numbers using scientific notation and introduce the concept of significant figures.
3. Students will multiply and divide small and large numbers using scientific notation.
4. Students will continue multiplying and dividing small and large numbers using scientific notation.

Content Notes for Lecture and Discussion

In this unit's Homework Assignment, students are asked to calculate the number of atoms that could fit into an object the size of our solar system. This calculation would be difficult, to say the least, using standard mathematical algorithms which are inadequate for handling numbers that describe phenomena at the atomic and cosmic scales. The diameter of an atom can be measured in Angström units where one Angström equals 0.00000001 cm (or 1×10^{-8} cm = 1×10^{-10} meters). At the opposite extreme are immense cosmic objects like stars, galaxies, and quasars. The star Betelgeuse, flickering red at the right shoulder of the constellation Orion, is one such common cosmic object. Betelgeuse is a red supergiant, 310 light-years from earth, measuring 500 times the diameter of our sun. Betelgeuse would fill our Solar System to the orbit of Jupiter. To help them complete their homework calculation, students are informed that an atom is one millionth-trillionth-trillionth of a cubic meter (1×10^{-30} m^3). The solar system is one thousand-trillion-trillion cubic meters in volume (1×10^{27} m^3). **Scientific notation** makes the solution to this problem easy to find.

After previewing the homework assignment at the start of Lesson #1, ask students to calculate the answer to Homework Problem #1 using standard division. Discuss the problems associated with this calculation: Was a calculator able to handle the problem? If students didn't use a calculator, did they enjoy lining up all those zeros? Obviously, trying to find the number of atoms that could fit into an object the same volume as our solar system is cumbersome using standard mathematical procedures. Scientific notation lends brevity to standard mathematical operations.

To describe phenomena on the atomic and cosmic scales, scientific notation employs base 10 raised or lowered to larger or smaller numbers using exponents. Using base 10 is consistent with the Metric System as all units of measure in metrics are multiples of 10. Average students will, at first, be uncomfortable working with exponents but, with practice, will find it preferable to standard math when dealing with very small and large numbers. Advanced students can be introduced to the concept of a **logarithm**. A logarithm is an exponent. A logarithm is defined as the power to which a number must be raised to give a particular value. The value "100" is "10 to the 2nd power" or "10^2" (i.e., 100 = 1×10^2). The logarithm of "100" is "2" in the base "10" (i.e., $\log_{10} 100 = 2$).

After attempting to do Homework Problem #1 at the start of Lesson #1, have students use a metric ruler to measure the length of a very small object like the graphite tip of a pencil or the metal point of a pen. Have students calculate how many lengths of the pencil or pen tip would reach to 1 centimeter. Then ask: "How many pencil (or pen) tips would it take to reach from the earth to the moon?" Inform students that the moon is about 390,000 kilometers from the earth (about 242,000 miles). Give students a moment to brainstorm this problem and, if possible, come up with an answer. Explain the basic strategy of scientific notation described in the Fact Sheet.

PS4 Content Notes *(cont'd)*

After making sure that units of measure are the same for both large and small objects (i.e., the distance to the moon and the length of the pencil tip, respectively), scientists ignore the 9 zeros in 390,000,000,000 millimeters and divide 390 by the length in millimeters of their pencil tip. They then tack the 9 zeros back onto their answer! In scientific notation, scientists "set aside" the zeros and work with smaller numbers and tack on the zeros later. Show students that large numbers are written as multiples of 10 in scientific notation: $100 = 10 \times 10$; $1{,}000 = 10 \times 10 \times 10$; $1{,}000{,}000 = 10 \times 10 \times 10 \times 10 \times 10 \times 10$. A similar strategy is used to write very small numbers: $1/10 = 1 \div 10$; $1/100 = 1 \div 10 \div 10$; $1/1{,}000 = 1 \div 10 \div 10 \div 10$. Have them refer to Chart #1 to see how scientists write large and small numbers using base 10. This unit can also include an introduction to the concept of significant figures which sets the limit on how accurately things can be measured. If someone reports the length of a pencil line to be 8.23764 cm, you'd guess they used a very accurate measuring device. A wooden metric ruler may be accurate to a half millimeter or so—say 8.25 cm—but not 8.23764 cm. The estimated length of 8.25 cm has three significant figures. The number of significant digits in the solution of any calculation is the same as the number of digits in the value with the fewest significant digits (i.e., the least accurate measurement).

Journal Sheets #1, #2, and #3 contain step-by-step instructions on how to add and subtract, multiply and divide small and large numbers using scientific notation. Throughout the week, brainstorm a variety of questions the students would like to answer that would require the use of scientific notation. Assist students in completing a variety of calculations based on real measurments made on common classroom objects. For example: If an atom is 0.00000001 cm, how many atoms laid end to end would fit along the edge of their notebook? If there are a billion billion galaxies in the universe (a conservative estimate), and a billion stars per galaxy, how may stars are there in the universe? By the end of the week, they should be able to find the answer to a host of questions requiring calculations using very small and very large numbers.

ANSWERS TO THE HOMEWORK PROBLEMS

1. 1,000,000,000,000,000,000,000,000,000 m^3 = 1×10^{27}
   ÷ 0.000000000000000000000000000001 m^3 = 1×10^{-30}
   let's not bother to write it = 1×10^{57} atoms

2. 1mm = $0.001\text{m} = 10^{-3}$ m
   1mm^3 = $0.001\text{m} \times 0.001\text{m} \times 0.001\text{m} = 10^{-3}\text{m} \times 10^{-3}\text{m} \times 10^{-3}\text{m} = 10^{-9}\text{m}^3$
   Earth radius (r) = 6,000km
   = $6 \times 10^3\text{km} \times 10^3\text{m/km} = 6 \times 10^6\text{m}$
   Earth volume (V) = $1.33\ \pi\ r^3$ (where $\pi = 3.14$)
   = $1.33 \times 3.14 \times 6 \times 10^6\text{m} \times 6 \times 10^6\text{m} \times 6 \times 10^6\text{m} \approx 902 \times 10^{18}\text{m}^3$
   Grains of sands in one earth volume = $902 \times 10^{18}\text{m}^3 \div 10^{-9}\text{m}^3$ per sand grain = 902×10^{27} grains of sand
   = 902,000,000,000,000,000,000,000,000,000 grains of sand

ANSWERS TO THE END-OF-THE-WEEK REVIEW QUIZ

1. D
2. I
3. M
4. K
5. O
6. C
7. F
8. N
9. E
10. H
11. L
12. A
13. G
14. J
15. B
16. 6.42×10^3
17. 0.00135
18. $20.8 \times 10^9 = 20{,}800{,}000{,}000$
19. $2.0 \times 10^6 = 2{,}000{,}000$
20. $6 \times 10^1 \times 6 \times 10^1 \times 2.4 \times 10^1 \times 3.6525 \times 10^2 \times 3 \times 10^5 = 946.728 \times 10^{10} \approx 9.5 \times 10^{12}$km

Name: ______________________ Period:________ Date: ___/___/___

PS4 Fact Sheet

MICRO TO MACRO: INTRODUCTION TO SCIENTIFIC NOTATION

CLASSWORK AGENDA FOR THE WEEK

(1) Explain the meaning of base 10 used in scientific notation.
(2) Write small and large numbers in scientific notation.
(3) Add and subtract small and large numbers using scientific notation.
(4) Multiply and divide small and large numbers using scientific notation.

Imagine measuring the length of a very small object like a grain of rice. Using a metric ruler, you might measure the length of the grain of rice to be 5 millimeters. How many grains of rice would reach to 1 centimeter? You know that 10 millimeters equals 1 centimeter, so, you would need 2 grains of rice to reach to 1 centimeter. That is . . .

$$\begin{aligned} 1 \text{ cm} &= 10 \text{ mm} \\ &\div \underline{5 \text{ mm per grain of rice}} \\ &= 2 \text{ grains of rice} \end{aligned}$$

Now, how about this question: How many grains of rice would it take to reach from the earth to the moon? The moon is about 390,000 kilometers from the earth (about 242,000 miles). To answer this question you would need to change 390,000 kilometers to 390,000,000,000 (390 billion) millimeters. Remember that each grain of rice is measured in millimeters and there are 1,000,000 millimeters in 1 kilometer. After that, you would have to divide 390,000,000,000 millimeters by 5 millimeters which is the length of each grain of rice. A shorthand way to get the correct answer would be to do the following. Ignore the 9 zeros after the number 390 and divide 390 by 5 to get 78. Then tack the 9 zeros back onto your answer! It would take 78,000,000,000 (78 billion) grains of rice to reach from the earth to the moon. Most calculators can't handle numbers in the billions or trillions. You can probably imagine how difficult it would be to do this kind of calculation using the longhand method. For this reason, scientists arrive at answers using the kind of strategy described above. They "set aside" the zeros and work with smaller numbers, then tack on the zeros later. They use a method called **scientific notation**.

In scientific notation, large numbers are written as multiples of 10. After all, $100 = 10 \times 10$; $1{,}000 = 10 \times 10 \times 10$; $1{,}000{,}000 = 10 \times 10 \times 10 \times 10 \times 10 \times 10$. A similar strategy is used to write very small numbers. Because, $1/10 = 1 \div 10$; $1/100 = 1 \div 10 \div 10$; $1/1{,}000 = 1 \div 10 \div 10 \div 10$. Study Chart #1 to see how scientists write large and small numbers using the basic number 10 called **base 10**.

CHART #1: Writing numbers in scientific notation using base 10.

number	standard notation	scientific notion
one trillion	1,000,000,000,000	1×10^{12}
one billion	1,000,000,000	1×10^{9}
one million	1,000,000	1×10^{6}
one hundred-thousand	100,000	1×10^{5}
ten thousand	10,000	1×10^{4}
one thousand	1,000	1×10^{3}
one hundred	100	1×10^{2}
ten	10	1×10^{1}
one	1	1×10^{0}
one tenth	0.1	1×10^{-1}
one hundredth	0.01	1×10^{-2}
one thousandth	0.001	1×10^{-3}
one hundred-thousandth	0.0001	1×10^{-4}
one millionth	0.00001	1×10^{-5}
one billionth	0.000000001	1×10^{-9}
one trillionth	0.000000000001	1×10^{-12}

In Chart #1, the small number written to the upper right of each base 10 in the scientific notation column is called an **exponent**. So, 1×10^2 (read "one times ten to the second power") means that the number 1 is *multiplied* by 10 twice ($1\times10\times10$). Multiplying $1\times10\times10$ gives 100. **Negative exponents** used to write numbers less than 1 mean that 1 is divided by 10. So, 1×10^{-2} means that the number 1 is divided by 10 twice ($1\div10\div10$) to get 0.01 or 1/100. Using scientific notation to add, subtract, multiply, and divide very small or large numbers is made easy by the fact that any number can be "substituted" for the number 1. For example: 3,450,000 = 3.45×10^6 (or $3.45\times10\times10\times10\times10\times10\times10$). To learn how to add and subtract, multiply and divide using scientific notation, follow the directions on your Journal Sheets.

Homework Directions

Solve the following problems using scientific notation.

1. An atom is a mere millionth-trillionth-trillionth of a cubic meter in volume (0.0000000000000 00000000000000001 cubic meter). The solar system is a thousand-trillion-trillion cubic meters in volume (1,000,000,000,000,000,000,000,000,000). How many atoms could fit, tightly packed, into an object the size of the solar system?

   Assignment due: __________

2. A grain of sand is about 1 cubic millimeter in volume. The earth is 6,000 kilometers in radius. How many grains of sand could fit inside an object the size of the earth? HINT: The volume of any sphere can be calculated using the following formula: $\mathbf{V=1.33\pi r^3}$ where r = the radius of the sphere and $\pi = 3.14$).

   Assignment due: __________

______________________ Student's Signature

______________________ Parent's Signature

___/___/___ Date

PS4 Lesson #1

MICRO TO MACRO: INTRODUCTION TO SCIENTIFIC NOTATION

Work Date: ___/___/___

LESSON OBJECTIVE

Students will explain the meaning of base 10 and write small and large numbers in scientific notation.

Classroom Activities

On Your Mark!

Preview the homework assignment on the Fact Sheet and give students several minutes to try calculating the answer to Homework Problem 1. Explain that scientific notation is a "shortcut" method scientists use to solve such problems.

Get Set!

Show students that large numbers can be written as multiples of 10 in scientific notation: 100 = 10 × 10; 1,000 = 10 × 10 × 10; 1,000,000 = 10 × 10 × 10 × 10 × 10 × 10. A similar strategy is used to write very small numbers: 1/10 = 1 ÷ 10; 1/100 = 1 ÷ 10 ÷ 10; 1/1,000 = 1 ÷ 10 ÷ 10 ÷ 10. Have students refer to Chart #1 to see how scientists write large and small numbers using base 10. Write standard numbers on the board and have students record on their Journal Sheet #1 how those numbers are written in scientific notation. Demonstrate how numbers can be written in scientific notation. (i.e., 1,243 = 1.243×10^3).

Go!

Have students use a metric ruler to measure the length of a very small object like the graphite tip of a pencil or the metal point of a pen. Have them calculate how many lengths of the pencil or pen tip would reach to 1 centimeter? Ask, "How many pencil (or pen) tips would it take to reach from the earth to the moon? The moon is about 390,000 kilometers from the earth (about 242,000 miles). Give students a moment to brainstorm this problem. Have students practice writing numbers in scientific notation and have students practice translating them back into standard numbers.

Materials

metric rulers, common objects

Name: ______________________________ Period:________ Date: ____/____/____

PS4 Journal Sheet #1

MICRO TO MACRO: INTRODUCTION TO SCIENTIFIC NOTATION

How to write numbers in scientific notation

Step#1

Move the decimal point so that there are only one or two digits to the left of the decimal point. Count the number of places you moved the decimal.

If the original number is greater than one . . .

Step#2A

. . . write "×10" to the right of your decimal number with a positive exponent equal to the number of places you counted.

If the original number is less than one . . .

Step#2B

. . . write "×10" to the right of your decimal number with a negative exponent equal to the number of places you counted.

Example A:

3,054,000

Step#1

3.054 can be written after moving the decimal 6 places to the left.

Step#2

3.054×10^6 is the answer because the original number is greater than one.

Example B:

0.000257

Step#1

2.57 can be written after moving the decimal 4 places to the right.

Step#2

2.57×10^{-4} is the answer because the original number is less than one.

PS4 **Lesson #2**

MICRO TO MACRO: INTRODUCTION TO SCIENTIFIC NOTATION

Work Date: ____/____/____

LESSON OBJECTIVE

Students will add and subtract small and large numbers using scientific notation and introduce the concept of significant figures.

Classroom Activities

On Your Mark!

Explain the step-by-step instructions on how to add and subtract small and large numbers using scientific notation that appears on the Journal Sheet #2. Brainstorm a variety of questions that students might like to answer that would require the use of scientific notation. Explain that scientists are careful not to overestimate the accuracy of the tools they use. Introduce the concept of significant figures and explain that calculations can only be as accurate as our measuring tools. If someone reports the length of a pencil line to be 8.23764 cm, they must have used a very accurate measuring device. A wooden metric ruler may be accurate to a half millimeter or so—say 8.25 cm—but not 8.23764 cm. The estimated length of 8.25 cm has three significant figures. The number of significant digits in the solution of any calculation is the same as the number of digits in the value with the fewest significant digits (i.e., the least accurate measurement).

Get Set!

Review how to change standard numbers into scientific notation and give several examples on the board demonstrating how to add and subtract numbers in scientific notation. Have students copy the examples on Journal Sheet #2.

Go!

Have students answer the questions posed during class discussion in cooperative activity.

Materials

metric rulers, common objects

Name: ______________________ Period:________ Date: ____/____/____

PS4 Journal Sheet #2

MICRO TO MACRO: INTRODUCTION TO SCIENTIFIC NOTATION

How to add and subtract numbers in scientific notation

Step#1

Move the decimal point so that all numbers have the same exponent after base 10.

Step#2

Line up decimal places and add or subtract the decimal portion of the number as you would normally. Bring down the base 10 with its exponent.

Step#3

Leave the answer in scientific notation or change it back by moving the decimal as indicated by the base 10 exponent (i.e., to a number greater than 1 if the exponent is positive or a number less than 1 if the exponent is negative).

Example A:

$$
\begin{aligned}
& 2.9\times10^4 = 2.900\times10^4 \\
+ \; & \underline{3.6\times10^2} = \underline{0.036\times10^4} \\
& \qquad\qquad = 2.936\times10^4
\end{aligned}
$$

Example B:

$$
\begin{aligned}
& 2.9\times10^4 = 2.900\times10^4 \\
- \; & \underline{3.6\times10^2} = \underline{0.036\times10^4} \\
& \qquad\qquad = 2.864\times10^4
\end{aligned}
$$

PS4 **Lesson #3**

MICRO TO MACRO: INTRODUCTION TO SCIENTIFIC NOTATION

Work Date: ____/____/____

LESSON OBJECTIVE

Students will multiply and divide small and large numbers using scientific notation.

Classroom Activities

On Your Mark!

Explain the step-by-step instructions on how to multiply and divide small and large numbers using scientific notation that appears on Journal Sheet #3. Brainstorm a variety of questions that students might like to answer that would require the use of scientific notation.

Get Set!

Give several examples that demonstrate how to multiply and divide numbers using scientific notation. Have students copy the examples on Journal Sheet #3.

Go!

Have students answer the questions posed during class discussion in cooperative activity. For example: If an atom is 0.00000001 cm, how many atoms laid end to end would fit along the edge of their notebook? Assist students in completing other calculations based on real measurements made on common classroom objects.

Materials

metric rulers, common objects

Name: ______________________ Period:________ Date: ___/___/___

PS4 Journal Sheet #3

MICRO TO MACRO: INTRODUCTION TO SCIENTIFIC NOTATION

How to multiply numbers in scientific notation

Step#1

Multiply the decimal part of the number as you would normally.

Step#2

Add the exponents for each base 10 to get your answer.

Example A:

$$\begin{array}{r} 3.7 \times 10^{8} \\ \times \ \underline{1.4 \times 10^{3}} \\ 5.18 \times 10^{11} \end{array}$$

How to divide numbers in scientific notation

Step#1

Divide the decimal part of the number as you would normally.

Step#2

Subtract the exponents for each base ten to get your answer.

Example A:

$$\begin{array}{r} 4.5 \times 10^{7} \\ \div \ \underline{0.5 \times 10^{5}} \\ 9.0 \times 10^{2} \end{array}$$

PS4 **Lesson #4**

MICRO TO MACRO: INTRODUCTION TO SCIENTIFIC NOTATION

Work Date: ____/____/____

LESSON OBJECTIVE

Students will continue multiplying and dividing small and large numbers using scientific notation.

Classroom Activities

On Your Mark!

Reinforce the skills learned in Lessons #1, #2 and #3 by brainstorming and solving problems that require the use of scientific notation.

Get Set!

Spend time with students who are still having difficulty working with base 10.

Go!

Assist students in completing calculations based on real measurements made on common classroom objects. Assist students in solving problems dealing with numbers on the atomic and cosmic scales (i.e., Homework Problems 1 and 2).

Materials

metric rulers, common objects

Name: ______________________________ Period:________ Date: ____/____/____

PS4 Journal Sheet #4

MICRO TO MACRO: INTRODUCTION TO SCIENTIFIC NOTATION

Name: ______________________ Period: ________ Date: ____/____/____

PS4 Review Quiz

Directions: Keep your eyes on your own work.
Read all directions and questions carefully.
THINK BEFORE YOU ANSWER!
Watch your spelling, be neat, and do the best you can.

CLASSWORK	(~40):	_______
HOMEWORK	(~20):	_______
CURRENT EVENT	(~10):	_______
TEST	(~30):	_______
TOTAL	(~100):	_______

(A ≥ 90, B ≥ 80, C ≥ 70, D ≥ 60, F < 60)

LETTER GRADE: _______

TEACHER'S COMMENTS: ______________________________________

MICRO TO MACRO: INTRODUCTION TO SCIENTIFIC NOTATION

MATCHING: Choose the letter of the number written in scientific notation on the right that is equal to the standard number on the left. *15 points*

_____	1. 472,000,000	(A)	4.72×10^{-7}
_____	2. 0.00472	(B)	4.72×10^{-4}
_____	3. 472	(C)	4.72×10^{5}
_____	4. 0.0000472	(D)	4.72×10^{8}
_____	5. 0.0000000472	(E)	4.72×10^{-6}
_____	6. 472,000	(F)	4.72×10^{3}
_____	7. 4,720	(G)	4.72×10^{7}
_____	8. 4,720,000	(H)	4.72×10^{-2}
_____	9. 0.00000472	(I)	4.72×10^{-3}
_____	10. 0.0472	(J)	4.72×10^{4}
_____	11. 4.72	(K)	4.72×10^{-5}
_____	12. 0.000000472	(L)	4.72×10^{0}
_____	13. 47,200,000	(M)	4.72×10^{2}
_____	14. 47,200	(N)	4.72×10^{6}
_____	15. 0.000472	(O)	4.72×10^{-8}

PROBLEMS

SHOW ALL OF YOUR WORK in solving problems #16 through #20. *15 points*

16. Write the number 6,420 in scientific notation.

17. Write 1.35×10^{-3} in standard notation.

18. Multiply 5.2×10^{6} by 4.0×10^{3} and write your answer in both scientific and standard notation.

19. Divide 6.0×10^{4} by 3.0×10^{-2} and write your answer in both scientific and standard notation.

20. Change each of the following numbers to scientific notation and use your calculator to multiply and discover the number of kilometers a light ray will travel in one year. Write your answer in scientific notation.

    60 seconds in one minute = ____________

    60 minutes in one hour = ____________

    24 hours in one day = ____________

    365.25 days in one year = ____________

    300,000 km per second = ____________ the speed of light

    Light travels ____________ kilometers in one year

____________________ Student's Signature

____________________ Parent's Signature

___/___/___ Date

PS5

MEASURING LENGTH, AREA, AND VOLUME

TEACHER'S CLASSWORK AGENDA AND CONTENT NOTES

Classwork Agenda for the Week

1. Students will draw a three-dimensional cube on a two-dimensional surface.
2. Students will calculate area from measures of length.
3. Students will calculate volume from measures of length.
4. Students will measure the volume of oddly shaped objects by water displacement.

Content Notes for Lecture and Discussion

Reading a professional journal article in physics or cosmology today is like taking a trip aboard the starship *Enterprise*. Such articles make the readers feel as though they've been warped into a world of pure science fiction. The "classical" view of space and time as perceived by Sir Isaac Newton (b. 1642; d. 1727) was changed forever by Albert Einstein (b. 1879; d. 1955) in the early decades of this century with the publication of his **Theories of Special and General Relativity**. Since Einstein, our concept of the universe has never been the same. Einstein and Niels Bohr (b. 1885, d. 1962), one of the first notable proponents of the new quantum physics, argued heatedly over the classical notions of time and space and how scientists might come to understand them. Modern physicists continue the argument today, decades after the deaths of these great scientists. Although middle school students need not be concerned with the details of these arguments, they ought to be aware that such a dialogue existed, and indeed still exists today among contemporary scientists.

Student awareness of standard units of measure, introduced in Plan Lesson P3, will help them to understand the idea that all units of measure for time, length, area, and volume are arbitrary units of measure. It is perfectly legitimate to base our understanding of the way nature behaves on arbitrary standards as long as we agree upon their meaning. A discussion of these facts will help students to appreciate that science is a cooperative activity that cannot be conducted in isolation. In addition, they will come to realize that science has its limitations. Scientists are a long way from knowing all there is to know about the universe. Students should be encouraged to consider that one day they may be among those scientists who evolve to a much clearer understanding of the natural world than that held by contemporary scientists.

In Lesson #1, open a discussion of the **three dimensions of space**. Define a **dimension** as "something through which you can move." Point out that we also move through "time" which is why it is considered the fourth dimension. Show students how to draw the three "spatial axes" shown on the Fact Sheet thereby representing a three-dimensional cube on a two-dimensional surface (i.e., Journal Sheet #1). Review basic units of measure in the Metric System (i.e., the centimeter and meter) and give students time to measure and compare the lengths of objects around the classroom.

In Lesson #2, begin a class discussion about the meaning of a flat surface or two-dimensional **plane**. Define an **area** as a confined region on a plane or flat surface. Demonstrate how to calculate the area of a square or rectangular flat surface by multiplying the lengths of the two perpendicular sides of any chosen surface. Explain that in metrics, area is measured in **square centimeters** or **square meters**. Allow students time to measure the lengths of the sides of rectangular surfaces around the classroom and to calculate the areas of those flat surfaces. Give them a selection of blocks or cardboard containers and direct them to find the surface areas of the individual sides as well as the total surface area of the objects (i.e., "How much wrapping paper would you need to cover the surface of this object?").

In Lesson #3, explain the meaning of the term **volume**. Define volume as the amount of space occupied by three dimensions. Explain that in metrics volume is measured in **cubic centimeters** or **cubic meters**. Demonstrate how to calculate the volume of a cube and give students time to measure the dimensions of blocks or cardboard boxes in order to determine the total volume of different objects.

In Lesson #4, demonstrate how to measure the volume of oddly shaped objects by water displacement. Explain that in metric units of measure, one **cubic decimeter** of volume is equal to one **liter**. One thousandth of a liter, a **milliliter**, is the same volume as one **cubic centimeter** of space.

ANSWERS TO THE END-OF-THE-WEEK REVIEW QUIZ

1. four
2. area
3. volume
4. true
5. true

Step#1: Fill a beaker or graduated cylinder with water and record the amount of water in the beaker.

Step#2: Completely submerge the oddly shaped object in the beaker and record the amount of water in the beaker.

Step#3: Subtract the amount of water you started with in Step #1 from the amount of water in the beaker in Step #2 to find the volume of the oddly shaped object.

PROBLEM

CRATE VOLUME

4 meters (length) × 3 meters (width) × 2 meters (height) = 24 cubic meters volume

STORAGE ROOM VOLUME

80 meters (length) × 10 meters (width) × 30 meters (height) = 24,000 cubic meters volume

FINAL ANSWER

1,000 crates will fit in the storage room.

Name: ______________________________ Period:________ Date: ____/____/____

PS5 Fact Sheet

MEASURING LENGTH, AREA, AND VOLUME

CLASSWORK AGENDA FOR THE WEEK

(1) Show how the three dimensions of space can be drawn on a two-dimensional surface.
(2) Show how area can be calculated from measures of length.
(3) Show how volume can be calculated from measures of length.
(4) Measure the volume of oddly shaped objects by water displacement.

As far as we know, the universe has four **dimensions**—three dimensions of **space** and a fourth dimension we call **time**. The three dimensions of space are all measures of **distance**. Measuring the distance to faraway objects is an ancient science. This science is a branch of mathematics called **geometry**. The prefix "geo" means "earth." The root word "metre" means "measure." Geometry has been used for thousands of years to calculate the distance to faraway objects and to help design structures like the giant pyramids of ancient Egypt.

The dimensions of any rectangular object can be found by calculation after measuring the object's **length**, **width**, and **height** using a ruler. Length, width, and height measure the distance between two points as drawn by a straight line. A three-dimensional, or 3-D, object can be drawn on a piece of paper (a two-dimensional surface) using "x," "y," and "z" axes as shown below.

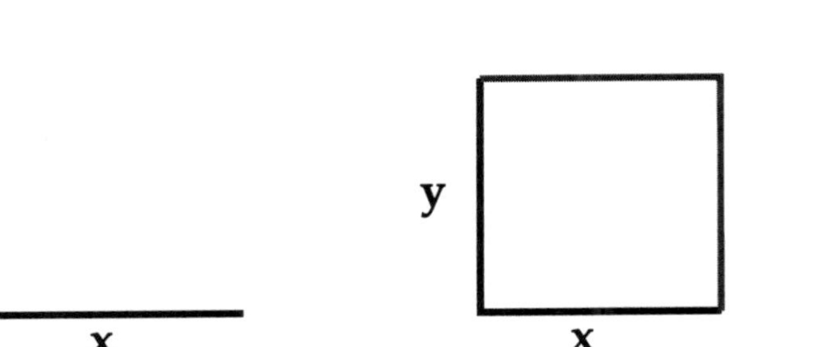

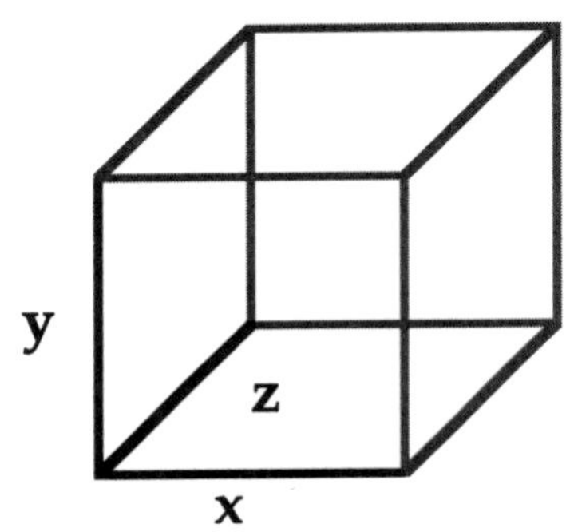

The "x" axis in each drawing shows the first dimension of space that we can call the "left to right" or "**horizontal axis**." The "y" axis in the second and third drawings shows a second dimension of space we can call the "up and down" or "**vertical axis**." The "z" axis in the last drawing shows the third dimension of space which appears to be going through the piece of paper from "front to back." That is the "z" axis which is drawn at an angle on the page. Drawing the "z axis" at an angle gives this flat, two-dimensional drawing the "3-D" look of a **cube**.

Every **surface**, or side, of a cube is a square: top, bottom, right and left, front and back sides. Opposite sides of a square are **parallel**. The sides connected at each "corner" of the cube are **perpendicular** to one another. Each side occupies a two-dimensional, flat **plane**.

An **area** is a confined region on a plane or flat surface. The area of a square or rectangular flat surface can be calculated by multiplying the lengths of the two perpendicular sides making up the surface (x "times" y, x "times" z, or y "times" z). In metrics, area is measured in **square centimeters** or **square meters**.

Volume is the amount of space occupied by three dimensions. The volume of a cube can be calculated by multiplying the lengths of the three perpendicular sides of the space (x "times" y "times" z). In metrics, volume is measured in **cubic centimeters** or **cubic meters**.

We can measure the volume of oddly shaped objects by "water displacement." Placing objects in a container of water causes the surface level of the water to rise. By measuring the amount of water displacement we can measure the volume of the oddly shaped object. In metric units of measure, one **cubic decimeter** of volume is equal to one **liter**. One thousandth of a liter, a **milliliter**, is the same as one **cubic centimeter** of space (or 1 ml = 1 cm^3).

A famous **physicist** named **Albert Einstein** (b., 1879; d., 1955), was one of the first scientists to realize that space and time cannot exist without one another. Space is dependent upon time and vice versa. Einstein called this new view of the universe the **space-time continuum**.

Homework Directions

1. Measure the length, width, and height of any rectangular room in your home in centimeters and meters using a metric ruler. Pick a simple room like a closet and not an oddly shaped room with vaulted ceilings or crooked walls.
2. Draw the room to scale labeling the floor, ceiling, and walls.
3. Calculate the volume of that room in cubic centimeters and cubic meters. SHOW ALL MEASURES AND CALCULATIONS!!

Assignment due: __________

_______________________ Student's Signature

_______________________ Parent's Signature

___/___/___ Date

PS5 Lesson #1

MEASURING LENGTH, AREA, AND VOLUME

Work Date: ____/____/____

LESSON OBJECTIVE

Students will draw a three-dimensional cube on a two-dimensional surface.

Classroom Activities

On Your Mark!

Open a discussion about the three dimensions of space. Define a dimension as "something through which you can move." Point out that we also move through "time" which is why it is considered the fourth dimension.

Get Set!

Show students how to draw the three "spatial axes" shown on the Fact Sheet on Journal Sheet #1. Review basic units of measure in the Metric System (i.e., the centimeter and meter).

Go!

Have students measure and compare the lengths of objects around the classroom in metric units of measure and record their measurements in Table A on Journal Sheet #1.

Materials

metric rulers, classroom desks, table tops, wooden blocks, empty cardboard containers (i.e., cereal boxes, shoe boxes, etc.)

Name: ______________________ Period: ________ Date: ___/___/___

PS5 Journal Sheet #1

MEASURING LENGTH, AREA, AND VOLUME

TABLE A			
object	length	width	height

PS5 **Lesson #2**

MEASURING LENGTH, AREA, AND VOLUME

Work Date: ____/____/____

LESSON OBJECTIVE

Students will calculate area from measures of length.

Classroom Activities

On Your Mark!

Begin a class discussion about the meaning of a flat surface or two-dimensional plane. Define an area as a confined region on a plane or flat surface.

Get Set!

Demonstrate how to calculate the area of a square or rectangular flat surface by multiplying the lengths of the two perpendicular sides of any chosen surface. Explain that in metrics, area is measured in square centimeters or square meters.

Go!

Have students measure the lengths of the sides of rectangular surfaces around the classroom and to calculate the areas of each surface. Have students record their data in Table B on Journal Sheet #2.

Materials

metric rulers, classroom desks, table tops, wooden blocks, empty cardboard containers (i.e., cereal boxes, shoe boxes, etc.)

Name: ______________________ Period:________ Date: ____/____/____

PS5 Journal Sheet #2

MEASURING LENGTH, AREA, AND VOLUME

TABLE B							
object	**side #1** (l × w)	**side #2** (l × h)	**side #3** (w × h)	**side #4** (l × w)	**side #5** (l × h)	**side #6** (w × h)	**total surface area**

PS5 Lesson #3

MEASURING LENGTH, AREA, AND VOLUME

Work Date: ___/___/___

LESSON OBJECTIVE

Students will calculate volume from measures of length.

Classroom Activities

On Your Mark!

Explain the meaning of the term volume. Define volume as the amount of space occupied by three dimensions. Explain that in metrics volume is measured in cubic centimeters or cubic meters. Refer to Figure A in Journal Sheet #3 and ask, "How many small blocks are there in the large block pictured in Figure A?"

Get Set!

Demonstrate how to calculate the volume of a cube.

Go!

Have students measure the dimensions of blocks or cardboard boxes to determine the total volume of different objects. Tell them to record their data on Table C in Journal Sheet #3.

Materials

metric rulers, wooden blocks, empty cardboard containers (i.e., cereal boxes, shoe boxes, etc.)

Name: ______________________________ Period:________ Date: ____/____/____

PS5 Journal Sheet #3

MEASURING LENGTH, AREA, AND VOLUME

FIGURE A
How many small cubes are in this block?

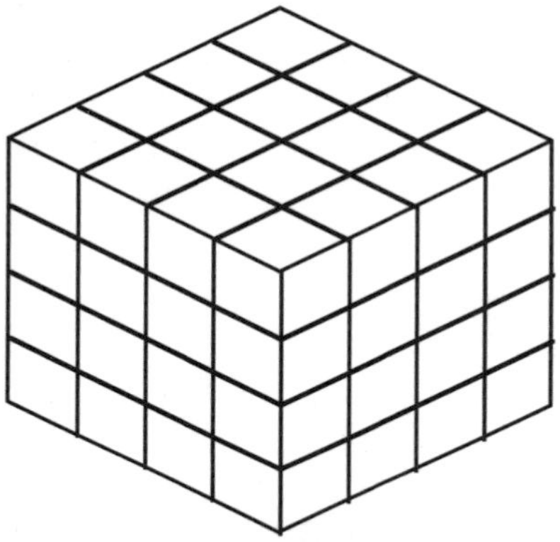

TABLE C					
object	**length**	**width**	**height**	**volume** (in cm^3)	**volume** (in m^3)

PS5 Lesson #4

MEASURING LENGTH, AREA, AND VOLUME

Work Date: ____/____/____

LESSON OBJECTIVE

Students will measure the volume of oddly shaped objects by water displacement.

Classroom Activities

On Your Mark!

Explain that in metric units of measure, one cubic decimeter of volume is equal to one liter. A liter is a commonly used metric unit of measure for liquid volume. One thousandth of a liter, a milliliter, is the same volume as one cubic centimeter of space. Begin a brief discussion about what happens when we submerge ourselves in a bathtub. Ask: "What happens to the level of water in the tub?" "How much does the water rise?" "Could the amount that the water rises be the same as our own body volume?" Explain that a similar method can be used to measure the volume of oddly shaped objects.

Get Set!

Demonstrate how to measure the volume of oddly shaped objects by water displacement. Remind students to record the amount of water in their beaker/graduated cylinder before they totally submerge their oddly shaped object. Tell them to subtract the reading they obtained *before* they submerged the object, from the reading obtained *following* total submersion of the object. The difference equals the volume of their object.

Go!

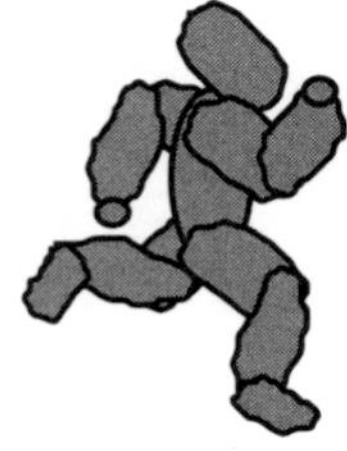

Have students measure the volume of wooden blocks and other oddly shaped objects by water displacement. Tell them to record their data on Table D in Journal Sheet #4. If they use the same wooden blocks as they used in Lesson #3, have them compare their "calculated measure" to their "water displacement measure." Ask: "Which of the two methods do you think was the most accurate?"

Materials

graduated cylinders, metric rulers, wooden blocks, empty cardboard containers (i.e., cereal boxes, shoe boxes, etc.)

Name: ______________________ Period:________ Date: ___/___/___

PS5 Journal Sheet #4

MEASURING LENGTH, AREA, AND VOLUME

TABLE D			
object	after submersion	before submersion	total volume

Name: ______________________ **Period:**________ **Date:** ___/___/___

PS5 Review Quiz

Directions: Keep your eyes on your own work.
Read all directions and questions carefully.
THINK BEFORE YOU ANSWER!
Watch your spelling, be neat, and do the best you can.

CLASSWORK	(~40): _______
HOMEWORK	(~20): _______
CURRENT EVENT	(~10): _______
TEST	(~30): _______
TOTAL	(~100): _______

(A ≥ 90, B ≥ 80, C ≥ 70, D ≥ 60, F < 60)

LETTER GRADE: _______

TEACHER'S COMMENTS: ____

____

MEASURING LENGTH, AREA, AND VOLUME

TRUE–FALSE FILL-IN: If the statement is true, write the word TRUE. If the statement is false, change the underlined word to make the statement true. *10 points*

_____________ 1. As far as we know, the universe has <u>three</u> dimensions.

_____________ 2. A(n) <u>volume</u> is a confined region on a flat surface.

_____________ 3. A(n) <u>area</u> is the amount of space occupied by three dimensions.

_____________ 4. The area of a rectangular surface can be calculated by multiplying the length of <u>two</u> perpendicular sides of that surface.

_____________ 5. The volume of a cube can be calculated by multiplying the length of <u>three</u> perpendicular sides of that surface.

SHORT ESSAY: Use the spaces below to describe how you would measure the volume of an oddly shaped object by water displacement. *10 points*

Step #1: ____

____

____

Step #2: ____

____

____

Step #3: ____

____

____

PS5 Review Quiz *(cont'd)*

PROBLEM: Show all math calculations. Use correct units of measure. Put answers in the spaces provided. *10 points*

How many crates with the dimensions listed on the left will fit into a storage room with the dimension listed on the right?

CRATE DIMENSIONS		STORAGE ROOM DIMENSIONS	
length:	4 meters	length:	80 meters
width:	3 meters	width:	10 meters
height:	2 meters	height:	30 meters

Crate Volume: _______________

Storage Room Volume: _______________

Final Answer: _______________

_______________________________ Student's Signature

_______________________________ Parent's Signature

____/____/____ Date

PS6

MEASURING MASS, VOLUME, AND DENSITY

TEACHER'S CLASSWORK AGENDA AND CONTENT NOTES

Classwork Agenda for the Week

1. Students will demonstrate the difference between mass and weight and construct a simple cardboard balance.
2. Students will measure the mass of a variety of "weightless" objects.
3. Students will measure the volume of oddly shaped objects.
4. Students will calculate the density of a variety of objects.

Content Notes for Lecture and Discussion

In everyday language, the terms **mass** and **weight** are frequently used interchangeably; to a scientist, there is an important distinction between the two. **Mass** is a measure of the amount of matter in an object as compared to a known standard amount of matter. Mass is an intrinsic property of an object and does not change regardless of the object's position or motion in space. **Weight**, on the other hand, is a measure of the force of gravity on a massive object. It *is not* an intrinsic property of the object itself. An object at the equator has less weight than an object at the North Pole because it is farther from the earth's center of gravity and under the influence of a centrifugal force directed away from that center. However, the mass of the object would be the same in both locations. The difference between the two concepts can be explained easily by prompting students to discuss the experiences of astronauts in space flight. The astronauts are weightless yet retain their body mass. Astronauts on the moon can hop around the lunar surface carrying themselves and a life support pack filled with instruments. The astronaut and pack weigh a total of 300 pounds on earth; on the moon, the astronaut and pack weigh a mere 50 pounds! Ask students to consider what a bathroom scale might read if they stood on it while jumping off a diving board. Do objects in freefall have weight? The answer is no—but they do have mass. If you sat on a bathroom scale during a ride on a roller coaster, would it give the same reading from the beginning to the end of the trip? Again, the answer is no. During the ride, your momentum would be changing constantly in favor of, and against, the pull of gravity. When a person says, "I'm trying to lose weight," is that really what they mean? If it was, they could have their wish quite easily by jumping off a cliff. They'd weigh nothing all the way down! What they really mean is, "I would like to lose mass—especially that part of me made of a substance called 'fat.' "

Discuss how a **spring scale** and a **balance** measure different things. Measuring the weight of an object using a spring scale could give a diversity of results depending on the position and motion of the object being measured. This is because the measure is determined by the force of gravity pulling on the spring. The reading on a balance gives a true measure of the amount of matter in an object. This is because the effect of gravity is cancelled by putting the object in equilibrium—in "balance"—with a known standard mass.

After reviewing the use of a balance and allowing students to measure the masses of a variety of objects, review the method for measuring the volume of oddly shaped objects by water displacement. Introduce the concept of **density** by discussing the meaning of the term. Define density as a measure of how "tightly" matter is packed inside an object. Explain that density is a **derived measure**—a calculated value based on two or more units of measure. A measure of density takes both mass and volume into account in order to arrive at a description of one physical property of an object. *Density is mass per unit volume.* Explain that speed is also a derived measure. A measure of speed takes both distance and time into account in order to arrive at a description of how fast an object is moving. In addition, explain to students that scientists make sure that large units of measure are always compared to large units of measure (i.e., the meter and kilogram) in the MKS or Meter–Kilogram System. Likewise, small units of measure are always compared to small units of measure in the CGS or Centimeter–Gram System. The solution to Homework Problem 2 requires that students are aware of this convention.

Throughout their study of physics, students will become familiar with a host of derived measures including **acceleration**, **momentum**, **force**, **work**, **energy**, and **power**.

ANSWERS TO THE HOMEWORK PROBLEMS

1. D = m ÷ v
   18.3 grams ÷ 6.1 cubic centimeters = 3 grams per cubic centimeter
2. D = m ÷ v
   0.124 kilograms = 124 grams
   1 milliliter = 1 cubic centimeter; so, 248 milliliters = 248 cubic centimeters
   124 grams ÷ 248 cubic centimeters = 0.5 grams per cubic centimeter
3. Ignore the fact that the metal alloy used to build the ship has a density of 8.0 grams per cubic centimeter. Consider the fact that the density of water is—by definition—1 gram per cubic centimeter (or 1 kilogram per liter). If the ship has a mass of 18,000 kilograms, then it must displace no less than 18,000 liters of water. In actuality, less sea water than fresh water need be displaced in order for the ship to stay afloat, since sea water is slightly more dense than fresh water.

Using their cardboard balance and pennies (= 3 grams each) students can find the mass of a small paper cup and add water to balance 2 pennies at a time (= 6 grams = 6 milliliters of water). Holding the cup up to the light after each balancing measure, students can draw a line at each level to create a "graduated cylinder."

ANSWERS TO THE END-OF-THE-WEEK REVIEW QUIZ

1. centimeter, meter (answers will vary)
2. mass
3. weight
4. gram, kilogram
5. true
6. time
7. true
8. true
9. true
10. true
11. D
12. A
13. E
14. B
15. C

PROBLEM

OBJECT MASS
5,400 kilograms

OBJECT VOLUME
6 meters (length) × 3 meters (width) × 3 meters (height) = 54 cubic meters

DENSITY OF THE OBJECT
D = m ÷ v
5,400 kilograms ÷ 54 cubic meters = 1,000 kilograms per cubic meter.
Note: This object is quite dense: 100 grams per cubic centimeter
or about 33 pennies packed into a space the size of a sugar cube.

Name: ______________________ Period: ________ Date: ___/___/___

PS6 Fact Sheet

MEASURING MASS, VOLUME, AND DENSITY

CLASSWORK AGENDA FOR THE WEEK

(1) Explain the difference between mass and weight.
(2) Measure the mass of a variety of "weightless" objects.
(3) Measure the volume of oddly shaped objects.
(4) Calculate the density of a variety of objects.

Space and time are not the only things that scientists can measure. As you learned in an earlier lesson, mass is another property of matter that can be measured directly. **Mass** is the amount of matter in an object and the unit of measure for mass is the gram. Water is the material standard used to find out how many grams of matter there are in an object. By definition, one cubic centimeter of water at 4° Celsius at sea level is equal to one **gram**.

In addition to space, time, and mass, there are a variety of other measureable quantities that can be calculated or "derived." Derived units of measure can be found by combining units of measure for distance, mass, and time. For example, speed is a **derived measure**. Speed describes how fast an object is moving. The speed of a moving object can be found by dividing the distance the object has travelled by the time it took to travel that distance. When we say an automobile is travelling at 70 kilometers per hour (kph) we are comparing a measure of distance (kilometers) with a measure of time (hours). Another derived unit of measure is weight. **Weight** is the force of gravity, or pull of the earth, exerted on an object that has mass. *Weight and mass do not have the same meaning*. Weight is a force. Mass is an amount of matter.

Another derived unit of measure is the gram per cubic centimeter. A **gram per cubic centimeter** measures the density of an object. **Density** describes how tightly matter is packed inside an object. Density **(D)** is the mass **(m)** per unit volume **(v)** of a substance. Twenty kilograms of cotton has the same mass as twenty kilograms of iron. However, the iron is normally more densely packed. It occupies a smaller amount of space than the fluffy cotton. Density can be calculated using the following mathematical formula:

$$\mathbf{D = m \div v}$$

There are many objects in our universe that are extremely dense. **Neutron stars** and **black holes** are two such objects. Scientists calculate that the matter inside a neutron star is so tightly packed that a chunk of such an object no larger than your thumb would weigh as much as a mountain. The matter at the center of a black hole is so tightly jammed together that everything inside the black hole must be "crushed" out of existence. You might think that the sun is a very dense object. But the sun is less dense on average than earth. Earth is the densest planetary object in our solar system. Earth is a rocky planet made mostly of solid iron. As large as it is, the sun is made mostly of gas like the giant outer planets, Jupiter, Saturn, Uranus, and Neptune.

Homework Directions

SHOW ALL MATHEMATICAL FORMULAS AND CALCULATIONS IN SOLVING PROBLEMS #1, #2, AND #3. BE SURE TO INCLUDE CORRECT UNITS OF MEASURE WITH YOUR ANSWER.

1. A piece of metal has a mass of 18.3 grams and a volume of 6.1 cubic centimeters. What is the density of the metal?
2. A block of wood has a mass of 0.124 kilograms and displaces 248 milliliters of water. What is the density of the block?
3. On average, the metal alloy used to build a ship has a density of 8.0 grams per cubic centimeter. The ship has a mass of 18,000 kilograms. How much water will the ship need to displace in order to stay afloat? *Hint:* An object will float in a fluid if it is less dense than the fluid into which it is placed. Find the density of water in Paragraph #1 of this Fact Sheet.

Assignment due: __________

USE THE CARDBOARD BALANCE YOU CONSTRUCTED IN CLASS TO CREATE A SMALL "GRADUATED" PAPER CUP (i.e., LIKE A GRADUATED CYLINDER) THAT WILL HELP YOU TO MEASURE THE VOLUME OF LIQUIDS IN 6 MILLILITER STEPS TO AT LEAST 60 MILLILITERS.

Assignment due: __________

______________________ Student's Signature

______________________ Parent's Signature

____/____/____ Date

PS6 Lesson #1

MEASURING MASS, VOLUME, AND DENSITY

Work Date: ____/____/____

LESSON OBJECTIVE

Students will demonstrate the difference between mass and weight and construct a simple cardboard balance.

Classroom Activities

On Your Mark!

Give students a few minutes to discuss the difference between mass and weight. Have them record their major points of discussion in Journal Sheet #1. Open a class discussion about the meaning of the two terms. Clarify the distinction between the two terms using the examples suggested in the Teacher's Classwork Agenda and Content Notes. Emphasize that mass is an intrinsic property of matter while weight is determined by the position and motion of an object in a gravitational field.

Get Set!

Show students how to use a spring scale before doing the first activity. Refer to Figure A in Journal Sheet #1. Preview the steps of the demonstrations outlined below in order to demonstrate that the weight of an object can change with a change in its position. The amount of matter in an object—its mass—remains unaltered by changes in position and motion. Distribute pieces of cardboard, scissors, metal nuts, and tape so students can construct their own cardboard balance.

Go!

Have students perform the following activity:

1. Suspend a small massive object from a spring scale and record its weight.
2. Lay the weight—still hooked to the end of the spring scale—flat on a table or desk.
3. Lift the table or desk at one end while holding on to the scale. Allow the weight to slip down the incline as the angle of the table inceases with respect to the floor.
4. Observe that the weight registering on the scale changes. Note that the weight of the object increases the greater the angle of the desk or table off the ground.

Give student time to construct the cardboard balance by following the directions in Figure B on Journal Sheet #1.

Materials

spring scales, cardboard, scissors, metric rulers, classroom desks or table tops, small massive objects (i.e., pieces of wood, nuts, bolts, rocks, etc.)

Name: ______________________ **Period:**________ **Date:** ___/___/___

PS6 Journal Sheet #1

MEASURING MASS, VOLUME, AND DENSITY

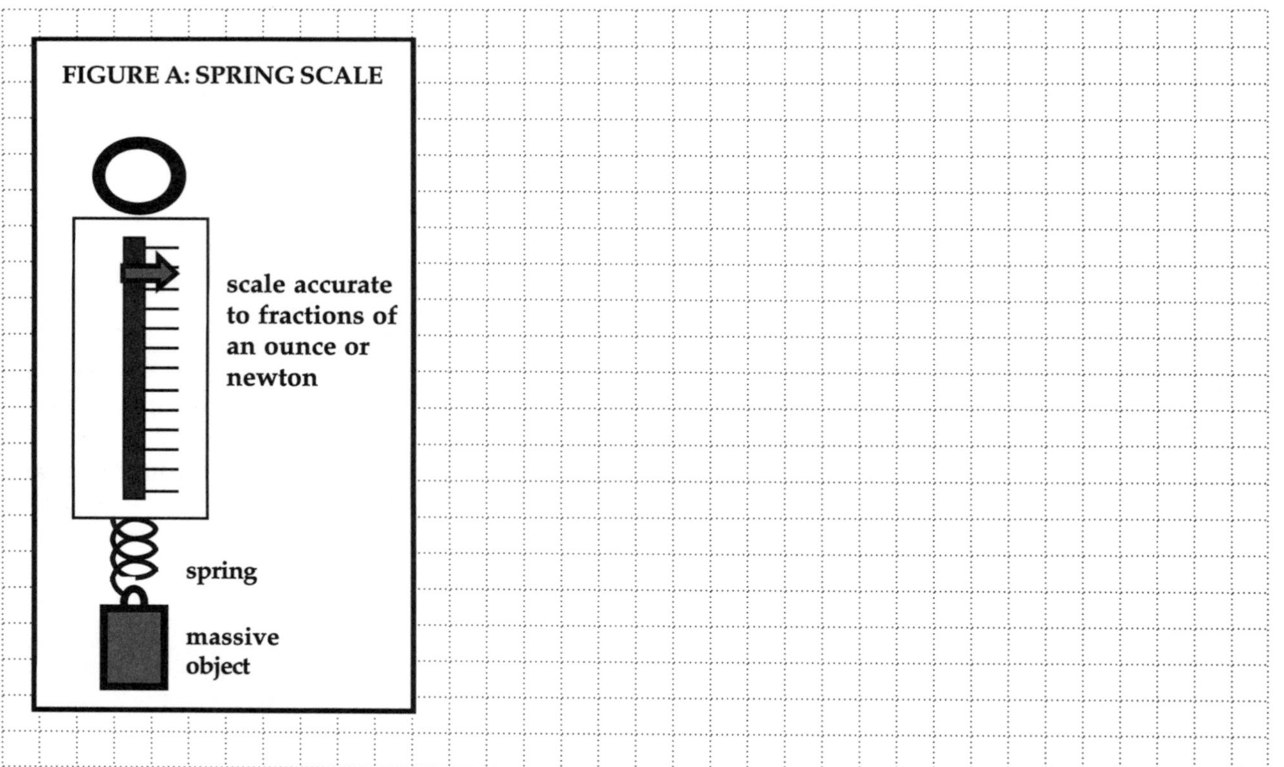

FIGURE B: A SIMPLE CARDBOARD BALANCE

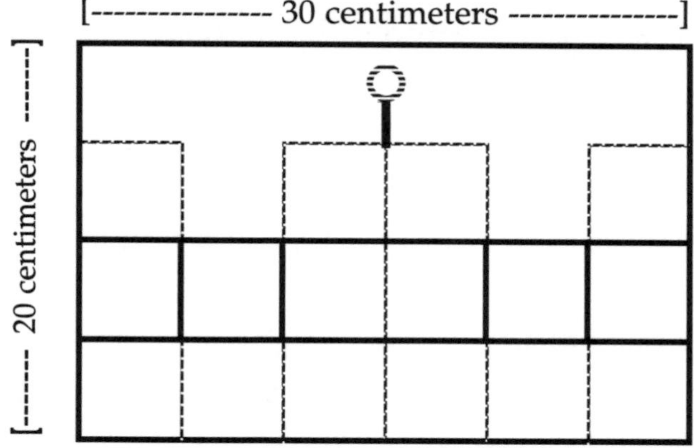

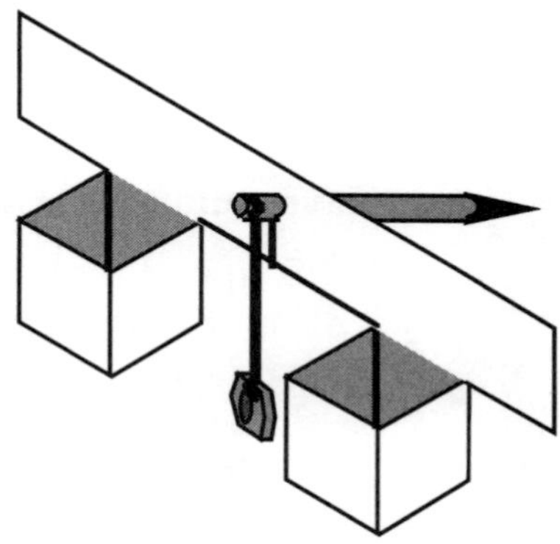

A) Cut out a piece of cardboard measuring 20 x 30 centimeters and draw squares measuring 5 x 5 centimeters, marking dotted and solid lines as shown.
B) Cut along all dotted lines, fold along all solid lines, and tape flaps to construct your balance as pictured on the right.
C) Punch out the hole at top-center so that a pencil will fit neatly through. Your balance should swing freely around the pencil.
D) Tie a small metal nut to a piece of string and suspend the string from the pencil. The string will hang straight down in line with the dark solid line drawn under the pencil hole when your object and standard mass units are in balance.
E) Use pennies as standard mass units (1 penny = 3 grams) to "weigh" your objects.

MEASURING MASS, VOLUME, AND DENSITY

Work Date: ____/____/____

LESSON OBJECTIVE

Students will measure the mass of a variety of "weightless" objects.

Classroom Activities

On Your Mark!

Review the distinction between mass and weight and explain that scientists measure the amount of matter in an object by comparing the object to a known standard mass. Define the gram as one cubic centimeter (or one milliliter) of pure water at 4° Celsius at sea level. Ask: "Why are temperature and elevation included in the definition of this standard?" Explain that the volume of water varies with changes in temperature and elevation, the latter due to changes in atmospheric pressure.

Get Set!

Review the use of a laboratory balance referring students to Figure C in Journal Sheet #2.

Go!

Have students measure and compare the masses of small massive objects in metric units using laboratory balances or their own cardboard balance. Have them record their measurements in Table A on Journal Sheet #2.

Materials

balances, student-made cardboard balances, pennies, small massive objects (i.e., pieces of wood, nuts, bolts, pebbles, marbles, etc.)

Name: ______________________ Period: ________ Date: ___/___/___

PS6 Journal Sheet #2

MEASURING MASS, VOLUME, AND DENSITY

TABLE A		
object	mass (grams)	mass (kilograms)

FIGURE C: DOUBLE BEAM BALANCE

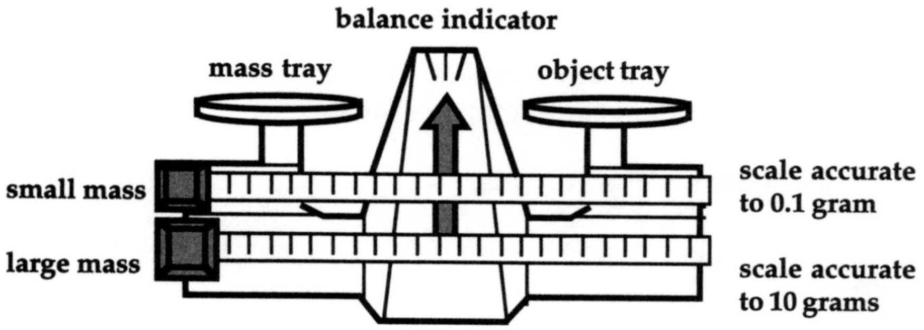

MEASURING MASS, VOLUME, AND DENSITY

Work Date: ____/____/____

LESSON OBJECTIVE

Students will measure the volume of oddly shaped objects.

Classroom Activities

On Your Mark!

Review the meaning of volume and how it can be "derived" by calculation or measured directly by water displacement.

Get Set!

Demonstrate how to calculate the volume of rectangular objects from measures of length, width, and height. Demonstrate how to measure the volume of an oddly shaped object by water displacement.

Go!

Have students calculate the volumes of small rectangular blocks of wood from measures of length, width, and height. Have them measure the volumes of the same blocks of wood by water displacement. Have them record the difference between their two measures and record the amount of their error. Have them record their data in Table B on Journal Sheet #3.

Materials

graduated cylinders, metric rulers, small massive objects (i.e., pieces of wood, nuts, bolts, rocks, etc.)

Name: ________________________ Period: ________ Date: ___/___/___

PS6 Journal Sheet #3

MEASURING MASS, VOLUME, AND DENSITY

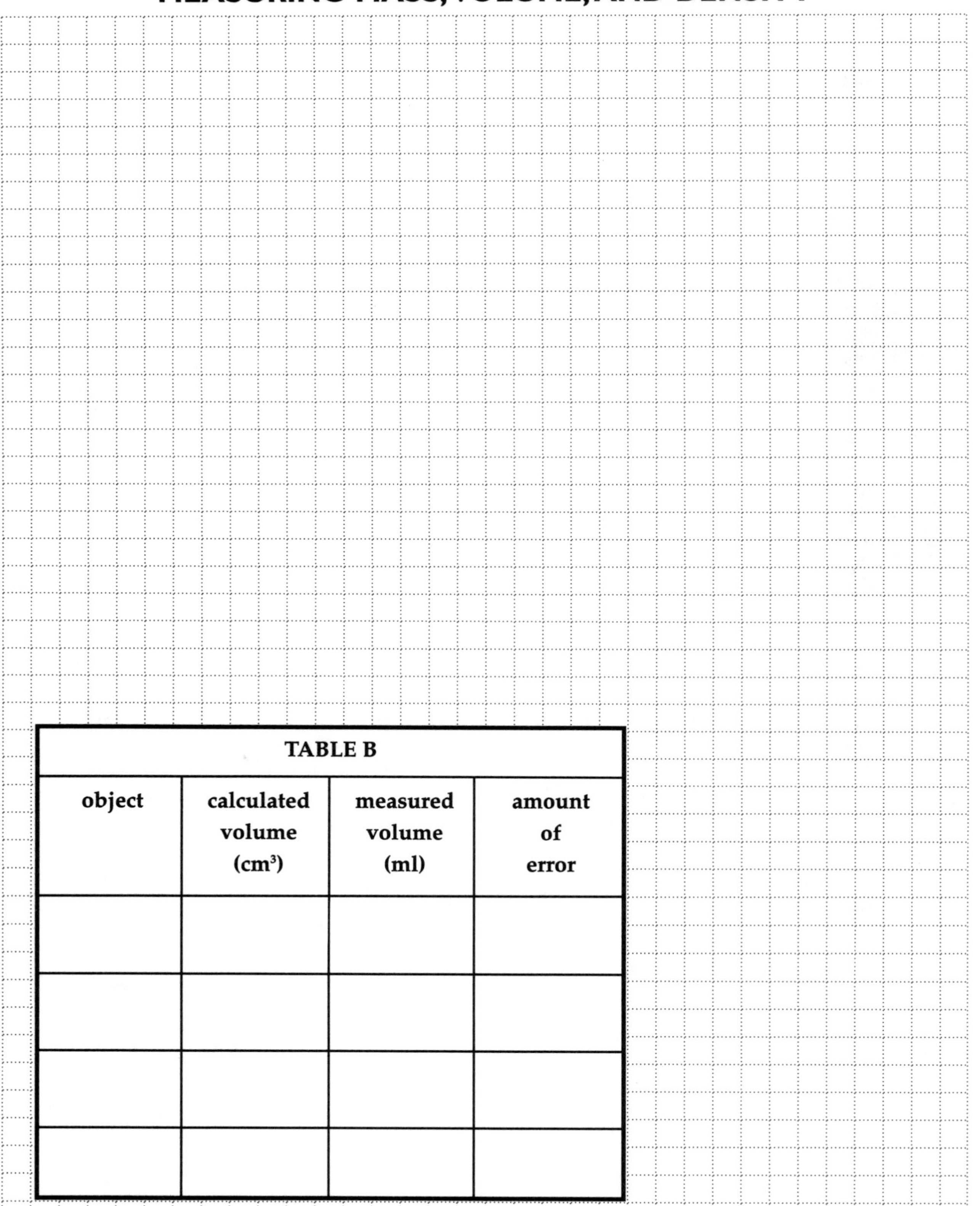

TABLE B			
object	calculated volume (cm^3)	measured volume (ml)	amount of error

MEASURING MASS, VOLUME, AND DENSITY

Work Date: ___/___/___

LESSON OBJECTIVE

Students will calculate the density of a variety of objects.

Classroom Activities

On Your Mark!

Explain why density is a derived measure. Point out that any measure of how tightly matter is packed into an object must take into account the amount of matter in the object (i.e., its mass) as well as the amount of space that the object occupies (i.e., its volume). Introduce the formula for density: $D = M \div V$; where D is density, M is mass, and V is volume. Have students refer to Figure C on Journal Sheet #4 to illustrate the meaning of the phrase "mass per unit volume."

Get Set!

Show students how to calculate the density of a variety of fictitious objects using the formula $D = M \div V$. They should already have some idea of how to do this from reading their Fact Sheet and trying the Homework Problems.

Go!

Have students measure the mass and volume of several objects and calculate the density of the objects. Have them record their data in Table C on Journal Sheet #4.

Materials

balances, graduated cylinders, small massive objects (i.e., pieces of wood, nuts, bolts, rocks, etc.)

Name: ______________________________ Period:________ Date: ____/____/____

PS6 Journal Sheet #4

MEASURING MASS, VOLUME, AND DENSITY

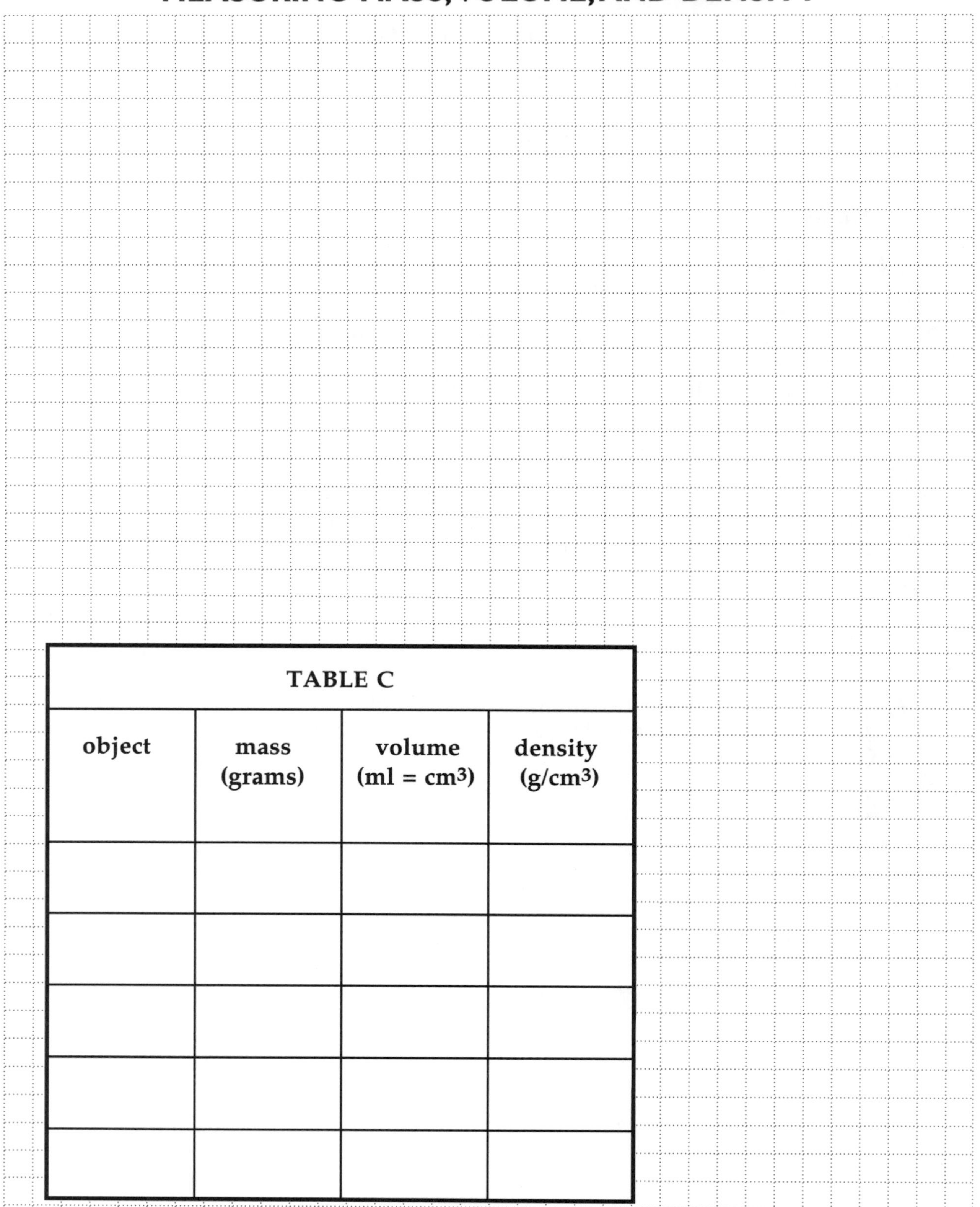

TABLE C

object	mass (grams)	volume (ml = cm^3)	density (g/cm^3)

Name: ______________________ **Period:**________ **Date:** ___/___/___

PS6 REVIEW QUIZ

Directions: Keep your eyes on your own work.
Read all directions and questions carefully.
THINK BEFORE YOU ANSWER!
Watch your spelling, be neat, and do the best you can.

CLASSWORK	(~40):	_______
HOMEWORK	(~20):	_______
CURRENT EVENT	(~10):	_______
TEST	(~30):	_______
TOTAL	(~100):	_______

(A ≥ 90, B ≥ 80, C ≥ 70, D ≥ 60, F < 60)

LETTER GRADE: _______

TEACHER'S COMMENTS: ____

MEASURING MASS, VOLUME, AND DENSITY

TRUE–FALSE FILL-IN: If the statement is true, write the word TRUE. If the statement is false, change the underlined word to make the statement true. *20 points*

_______________ 1. A measure for length in the metric system is the <u>yard</u>.

_______________ 2. <u>Weight</u> is the amount of matter in an object.

_______________ 3. <u>Mass</u> is the force of gravity exerted on an object.

_______________ 4. The unit of measure for mass is the <u>pound</u>.

_______________ 5. A gram is a cubic <u>centimeter</u> of pure water at 4° Celsius at sea level.

_______________ 6. When we say that a rocket is travelling at 17,500 "miles per hour," we are comparing a measure of distance with a measure of <u>speed</u>.

_______________ 7. A gram per cubic centimeter is a measure of <u>density</u>.

_______________ 8. <u>Density</u> is a measure of mass per unit volume.

_______________ 9. <u>Mass</u> is measured in grams.

_______________ 10. <u>Volume</u> is measured in cubic centimeters.

MATCHING: Choose the letter of the phrase on the right that best describes the unit of measure on the left. *5 points*

_____ 11. centimeter

_____ 12. gram

_____ 13. kilograms per cubic decimeter

_____ 14. square meter

_____ 15. cubic kilometer

(A) amount of matter in an object

(B) surface area

(C) three-dimensional space

(D) distance

(E) tightly packed matter

PROBLEM

Directions: Show all math calculations. Use correct units of measure. Put answers in the spaces provided!! *10 points*

What is the density of the object described below in kilograms per cubic meter?

OBJECT MASS
5,400 kilograms

OBJECT DIMENSIONS
length = 6 meters
width = 3 meters
height = 3 meters

OBJECT VOLUME: ____________

OBJECT DENSITY: ____________

______________________ Student's Signature

______________________ Parent's Signature

___/___/___ Date

PS7

SPEED, VELOCITY, AND ACCELERATION

TEACHER'S CLASSWORK AGENDA AND CONTENT NOTES

Classwork Agenda for the Week

1. Students will describe how speed is different than velocity.
2. Students will calculate the velocity of moving objects.
3. Students will graph the positions of objects accelerating and those moving at a constant velocity.
4. Students will measure and calculate the acceleration of moving objects.

Content Notes for Lecture and Discussion

The science of classical mechanics, the laws of motion and gravity as described by the seventeenth century mathematician-philosopher, Sir Isaac Newton (b. 1642; d. 1727), is founded upon the concepts of **force**, **motion**, and **energy**—all derived measures. The amount of energy that an object has depends upon its motion; so, physics, the study of the interaction between matter and energy, is, for the most part, the study of moving objects.

Motion can be described as either a **scalar** or **vector quantity**. A scalar quantity is a measure of magnitude alone. Readings on a bathroom scale are scalar measures of weight. Distance is also a scalar quantity. A vector, on the other hand, is a measure of both magnitude and direction.

Speed is a derived scalar quantity measuring the distance travelled by an object in a given amount of time. Speed is distance per unit time: $S = D \div t$, where S is speed, D is distance, and t is time. **Velocity** is a derived vector quantity that is calculated in the same manner as speed, keeping in mind the direction that the object is moving. **Acceleration** is a change in velocity: a change in either direction or speed in a given amount of time.

In Lessons #1 and #2, assist students in graphing, measuring, and calculating the speed and velocity of a variety of simple objects in motion. Work out problems on the board so that students can grasp the concept of speed and velocity as derived measures.

In Lesson #3, explain the difference between an object moving at a constant velocity and one that is accelerating. Students sometimes have difficulty conceiving of an object that accelerates at a constant rate while its velocity continues to change. To illustrate the fact that an object's velocity can change as it accelerates at a "constant rate," introduce the following example: Have them imagine a rock rolling down a hill. The rock rolls one meter in the first second, two meters in the second second, three meters in the third second, and so on. Ask: "How much farther is the rock rolling every second compared to the previous second?" Answer: "1 meter farther every second." The rate of acceleration of the rock is, therefore, 1 meter per second every second (m/s/s). Then ask: "What is the rock's speed at the end of the first, second, and third seconds?" Answer: "1 meter per second (1 m/s), two meters per second (2 m/s), and three meters per second (3 m/s), respectively. Point out that while the velocity of the rock increases, the "rate of accleration" remains the same [1 meter per second every second (m/s/s)].

PS7 Content Notes *(cont'd)*

Explain that acceleration is expressed in units of distance per unit of time "squared." Students' knowledge of fractions will help them to understand why acceleration is expressed in this unit of measure. Give the following example:

To solve the problem below

$$\frac{1}{2} \div \frac{3}{4} = ?$$

we can **invert** one of the fractions and multiply (i.e., numerator by numerator and denominator by denominator)

$$\frac{1}{2} \times \frac{4}{3} = \frac{4}{6} \text{ or } \frac{2}{3}$$

In physics, units of measure are treated in the same fashion as numbers. So . . .

$$\frac{\text{meters}}{\text{second}} \div \frac{\text{second}}{1} = \frac{\text{meters}}{\text{second}} \times \frac{1}{\text{second}} = \frac{\text{meters}}{\text{second}^2}$$

ANSWERS TO THE HOMEWORK PROBLEMS

1. S = D ÷ t
   S = 125 km ÷ 5 h = 25 kph during the first 5 hours
   D = S × t; so 25 kph × 2 hours after leaving the restaurant equals . . .
   Answer: 50 miles
2. Total distance jogged = 2 km + 3 km + 4 km + 3 km = 12 km
   S = D ÷ t
   S = 12 km ÷ 1 h = 12 kph
   Referring to the diagram to the right, one can see that the jogger would be 2 km west of his starting point at the end of his jog.

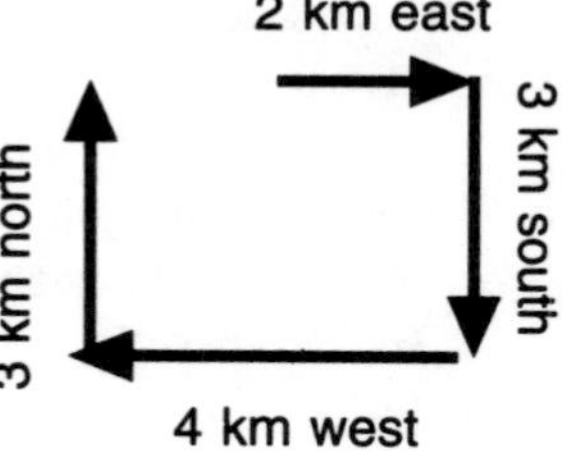

3. $a = (v_f - v_s) \div t$
   $a = (7 \text{ km/s} - 2 \text{ km/s}) \div 100 \text{ s} = 5 \text{ km/s} \div 100 \text{ s} = 0.05 \text{ km/s}^2$ or . . .
   50 meters per second every second

ANSWERS TO THE END-OF-THE-WEEK REVIEW QUIZ

1. true
2. true
3. direction
4. acceleration
5. speed or velocity
6. always
7. true
8. true
9. C
10. E
11. B
12. A
13. D

<u>PROBLEM #1</u>
S = D ÷ t
S = 500 km ÷ 10 h = 50 kph

<u>PROBLEM #2</u>
$a = (v_f - v_s) \div t$
$a = (5 \text{ m/s} - 35 \text{ m/s}) \div 5 \text{ s} = -30 \text{ km/s} \div 5 \text{ s} = -6 \text{ m/s}^2$
(NOTE: The negative acceleration indicates that the roller coaster is slowing or "decelerating.")

Name: ______________________________ Period:________ Date: ___/___/___

PS7 Fact Sheet

SPEED, VELOCITY, AND ACCELERATION

CLASSWORK AGENDA FOR THE WEEK

(1) Describe how speed is different than velocity.
(2) Calculate the velocity of moving objects.
(3) Compare objects moving at a constant velocity to those that are accelerating.
(4) Calculate the acceleration of moving objects.

In order to give a complete description of an object in motion, a scientist includes details about both the **speed** and **direction** of the object. Reporting an object's speed without giving its direction, or vice versa, would not be a complete physical description. A rapidly moving truck is of little concern to anyone except the people toward which the truck is moving! The term "velocity" refers to both the speed and the direction of a moving object.

Speed is a **scalar quantity**. A scalar quantity gives only the "amount" of something. **Velocity** is a **vector quantity**. It provides both the speed and direction of a moving object. Vectors are arrows that give "direction." Draw an arrow over the number indicating speed to show the velocity of a moving object.

$$\overrightarrow{50}\ \text{cm/s} = 50 \text{ centimeter per second to the right}$$

The speed **(s)** of a moving object is calculated by measuring the distance **(d)** through which the object travels and dividing that distance by the time **(t)** it took to travel that distance.

$$s = d \div t$$

The metric units of measure for speed are centimeters/second (**cm/s**), meters/second (**m/s**), or kilometers/second (km/s). The units of measure for velocity **(v)** are the same as that for speed. Velocity is calculated in the same way but includes a "directional arrow" to show the object's direction.

$$\vec{v} = d \div t$$

When an object changes its speed or direction it is accelerating. The **acceleration** (**a**) of a moving object is calculated by subtracting its starting velocity (v_s) from its final velocity (v_f), then dividing the difference by the time it takes to make that change. The unit of measure for acceleration can be expressed in cm/s/s (or **cm/s²**), m/s/s (or **m/s²**), or or km/s/s (**km/s²**).

$$\vec{a} = (v_f - v_s) \div t \quad \text{or} \quad \vec{a} = \Delta v \div t$$

The symbol Δ means "change."

The speed, velocity, and acceleration of an object are all important properties to a **physicist**. Each quantity determines the **energy** an object has. A small object moving at high velocity can have a lot more energy than a large object moving at low velocity.

Homework Directions

SHOW ALL MATHEMATICAL FORMULAS AND CALCULATIONS IN SOLVING PROBLEMS #1, #2, AND #3. BE SURE TO INCLUDE CORRECT UNITS OF MEASURE WITH YOUR ANSWER.

1. A man on a bicycle travels 125 kilometers in 5 hours. He stops for dinner at a restaurant along the road, then leaves. He averages the same speed for 2 more hours before reaching his destination. How far did he travel after leaving the restaurant?
2. A jogger runs east for 2 kilometers, south for 3 kilometers, west for 4 kilometers, and north for 3 kilometers. The entire run takes 1 hour. Express your answer in kilometers per hour and find the jogger's average speed for the entire trip. Where would the jogger be relative to his/her starting point at the end of the jog?
3. At 200 seconds after liftoff, a high-speed motion picture camera clocks a high speed-rocket moving at 2 km/s. At 300 seconds after liftoff, the rocket is moving at 7 km/s. How fast is the rocket accelerating into orbit during the elapsed time? How many meters in distance travelled is the rocket adding to the distance it travelled in the previous second?

Assignment due: __________

_______________________ Student's Signature

_______________________ Parent's Signature

___/___/___ Date

PS7 Lesson #1

SPEED, VELOCITY, AND ACCELERATION

Work Date: ____/____/____

LESSON OBJECTIVE

Students will describe how speed is different than velocity.

Classroom Activities

On Your Mark!

Begin the lesson by asking students to recall the last time they were a passenger in a moving vehicle. Ask them to write down in Journal Sheet #1 how fast they were moving and to include the "unit of measure" for speed. Ask them to discuss what is meant by that unit of measure and point out that "miles per hour" or "kilometers per hour" represents a comparison between distance and time. Write the formula for calculating speed on the board or overhead transparency: $S = D \div t$. Ask them to consider the direction in which they were moving. Point out that moving at a high rate of speed does not necessarily mean you will get to your destination sooner than if you were moving at a low rate of speed. You could be moving "away" from your destination! Ask them to consider the example given in their Fact Sheet that refers to a truck moving toward, as opposed to away from, a pedestrian. Explain that velocity is a more accurate description of an object's motion because it takes both speed and direction into account. Define speed and velocity as scalar and vector quantities, respectively.

Get Set!

Show students how to calculate the speed of a number of moving objects using the formula $S = D \div t$. Show them how to set up a graph on Journal Sheet #1 that will help them to record how far, and to calculate how fast, an object travels during different spans of time.

Go!

Have students graph the information shown in Table A on Journal Sheet #1. After completing the graph, give them time to copy and answer the following questions:

(A) How far did the object travel during the first 3 hours of its journey?
(B) What was the average speed of the object between 11 AM and 1 PM?
(C) How far did the object travel from 10 AM to 12 noon?
(D) What was the object's average speed from 8 AM to 5 PM?

Answers: (A) 12 km; (B) 6 kph; (C) 10 km; (D) 4 kph

Materials

metric rulers

Name: ______________________ Period: ________ Date: ___/___/___

PS7 JOURNAL SHEET #1

SPEED, VELOCITY, AND ACCELERATION

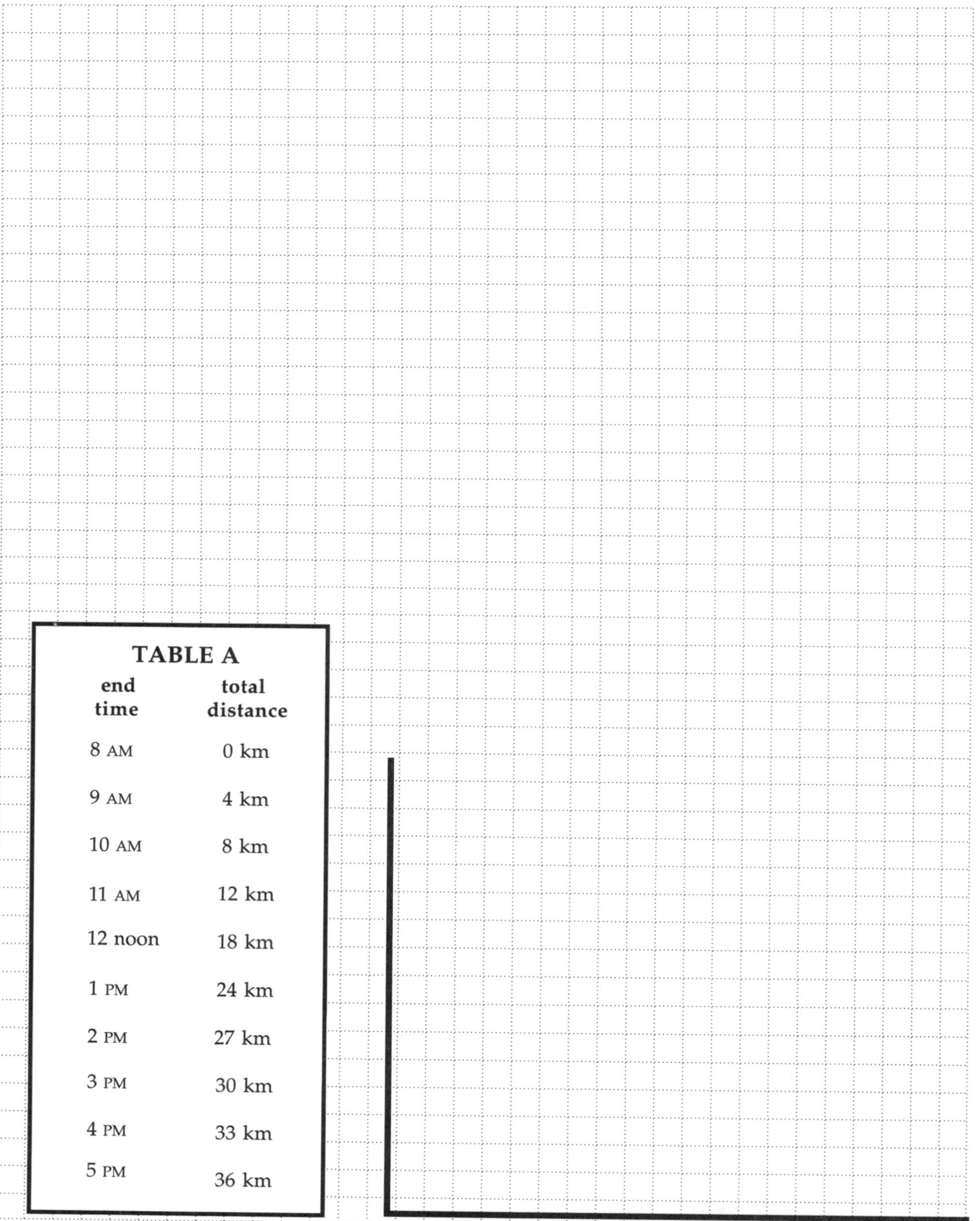

TABLE A

end time	total distance
8 AM	0 km
9 AM	4 km
10 AM	8 km
11 AM	12 km
12 noon	18 km
1 PM	24 km
2 PM	27 km
3 PM	30 km
4 PM	33 km
5 PM	36 km

PS7 Lesson #2

SPEED, VELOCITY, AND ACCELERATION

Work Date: ____/____/____

LESSON OBJECTIVE

Students will calculate the velocity of moving objects.

Classroom Activities

On Your Mark!

Write the formula for calculating speed on the board or overhead transparency: $S = D \div t$. Review how to use the formula referring back to Journal Sheet #1.

Get Set!

Show students how to set up the laboratory apparatus shown in Journal Sheet #2 to measure the speed of a rolling ball. If stopwatches are not available, show students how to use a pendulum (i.e., a nut suspended from a string hung over their desk or lab table) to obtain an arbitrary measure of elapsed time (i.e., speed = centimeters per swing in lieu of centimeters per second).

Go!

Have students practice timing the release and capture of the rolling ball, rounding off to the nearest swing if a pendulum is used as a timer. Have students complete Table B on Journal Sheet #2.

Materials

ring stand and clamps (or stacked books), metric rulers, balls (i.e., marbles, golf balls, ping pong balls, tennis balls, etc.), stopwatches (or string and metal nuts)

Name: ______________________ Period: ________ Date: ____/____/____

PS7 Journal Sheet #2

SPEED, VELOCITY, AND ACCELERATION

TABLE B					
distance (cm)	time (s)				average speed (cm/s)
	1st trial	2nd trial	3rd trial	avg. time(s)	

Time your rolling ball 3 times for each distance and calculate an average time before calculating average speed at each distance. Use five different distances from the top of the ramp.

FIGURE A

Tape 3 meter sticks together lengthwise to form a ramp for rolling objects. Use books in lieu of ring stands and clamps to adjust ramp height. Use a pendulum if stopwatches are unavailable.

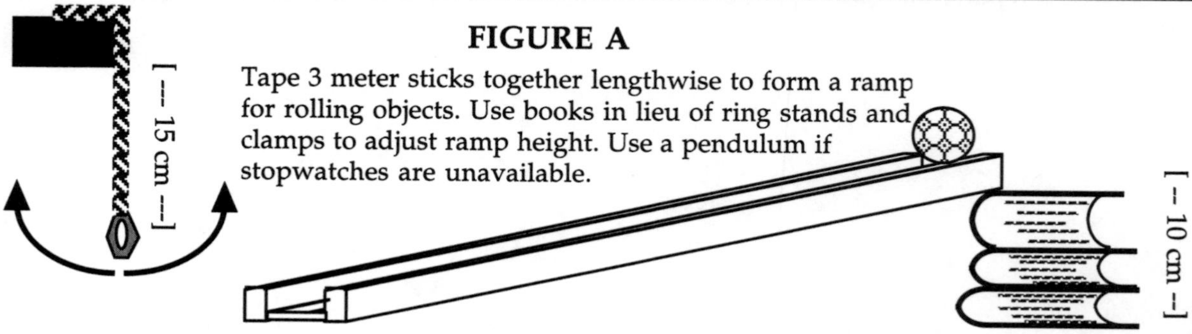

PS7 **Lesson #3**

SPEED, VELOCITY, AND ACCELERATION

Work Date: ____/____/____

LESSON OBJECTIVE

Students will graph the positions of objects accelerating and those moving at a constant velocity.

Classroom Activities

On Your Mark!

Briefly discuss the work of Galileo Galilei (b. 1564; d. 1642). In disagreement with the scholars of his time, Galileo experimented to show that all objects, regardless of their mass, accelerate toward the earth's surface at a constant rate. To demonstrate this point, he dropped objects of differing mass from the top of the Leaning Tower of Pisa and watched them hit the ground at the same time. Galileo's work laid the foundation for Sir Isaac Newton's (b. 1642; d. 1727) Law of Gravity. Gravity is an accelerating force, pulling objects toward the earth's surface at the rate of 9.8 meters per second every second ($9.8\ m/s^2$).

Get Set!

Work through several simple examples that will show students how to calculate the acceleration of a moving object using the formula $a = (v_f - v_s) \div t$. Show them how to set up a graph on Journal Sheet #3 that will help them to record how far an object moving at a constant velocity compares to one that is accelerating.

Go!

Have students graph the information shown in Table C on Journal Sheet #3. After completing the graph, give them time to copy and answer the following questions:

(A) Which object traveled farther in the first 3 hours: Object A or Object B?
(B) What was the average speed of Object A between 8 AM and 1 PM?
(C) What was the average speed of Object B between 8 AM and 1 PM?
(D) What was Object A's average velocity during the trip?
(E) What was Object B's rate of acceleration during the trip?

Answers: (A) both travelled the same distance; (B) 2 kph; (C) 3 kph; (D) 2 kph; (E) 1 km/h^2

Materials

metric rulers

Name: ______________________ Period:________ Date: ___/___/___

PS7 Journal Sheet #3

SPEED, VELOCITY, AND ACCELERATION

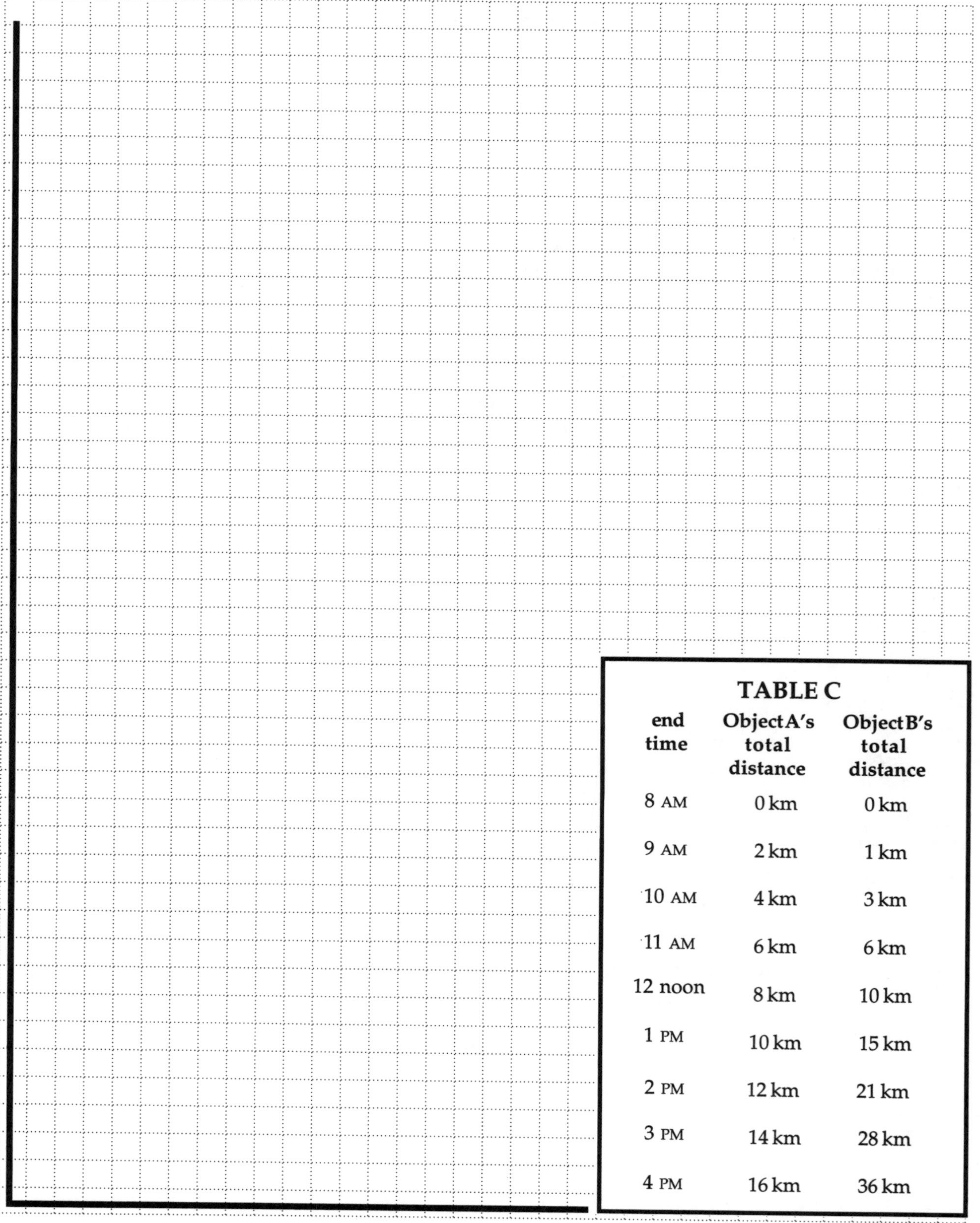

TABLE C

end time	Object A's total distance	Object B's total distance
8 AM	0 km	0 km
9 AM	2 km	1 km
10 AM	4 km	3 km
11 AM	6 km	6 km
12 noon	8 km	10 km
1 PM	10 km	15 km
2 PM	12 km	21 km
3 PM	14 km	28 km
4 PM	16 km	36 km

SPEED, VELOCITY, AND ACCELERATION

Work Date: ____/____/____

LESSON OBJECTIVE

Students will measure and calculate the acceleration of moving objects.

Classroom Activities

On Your Mark!

Explain that Galileo used a ramp to study the acceleration of falling objects, because free falling objects accelerated to the ground much too fast for him to measure. He didn't have a stopwatch or even a pendulum clock with which to time the fall or roll of objects. Instead, Galileo sang. As he sang, he marked the position of the rolling ball on each beat of the melody's rhythm. He then measured the distance the object had travelled between beats. In this manner, he discovered that the distance a falling object falls, or rolls down an incline, increases by the same amount with the passage of each regular unit of time. That is, the "change" in the object's velocity is constant. Refer students to their graph on Journal Sheet #3. If they completed the graph correctly, they can see how the object in that problem accelerated at a constant rate of 1 km/h^2. The velocity of the object changed by 1 km per hour every hour.

Get Set!

Work through several more examples that will show students how to calculate the acceleration of a moving object using the formula $a = (v_f - v_s) \div t$. Review how to set up the laboratory apparatus used in Lesson #2 to measure the acceleration of a rolling ball. If stopwatches are not available, review how to use a pendulum (i.e., a nut suspended from a string hung over their desk or lab table) to obtain an arbitrary measure of elapsed time (i.e., speed = centimeters per <u>swing</u> in lieu of centimeters per <u>second</u>).

Go!

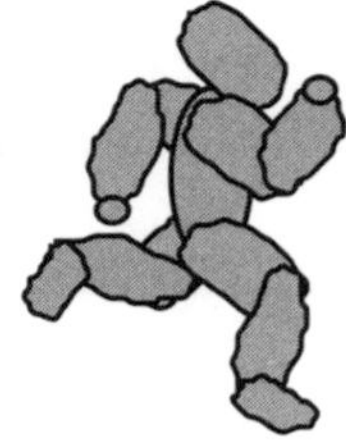

Have students practice timing the release and capture of the rolling ball, rounding off to the nearest swing if a pendulum is used as a timer. Have students complete Table D on Journal Sheet #4. If students perform this activity with a reasonable degree of accuracy, they should find that the rates of acceleration for all rolling objects, regardless of their mass, is the same. The rate of acceleration for an object rolling down from 25 cm to 50 cm is the same as the rate of acceleration for an object rolling down from 75 cm to 100 cm. Point out that while the rate of acceleration remained constant, the velocity, of course, continued to increase.

Materials

ring stand and clamps (or stacked books), metric rulers, balls (i.e., marbles, golf balls, ping pong balls, tennis balls, etc.), stopwatches (or string and metal nuts)

Name: ______________________ Period: ________ Date: ___/___/___

PS7 Journal Sheet #4

SPEED, VELOCITY, AND ACCELERATION

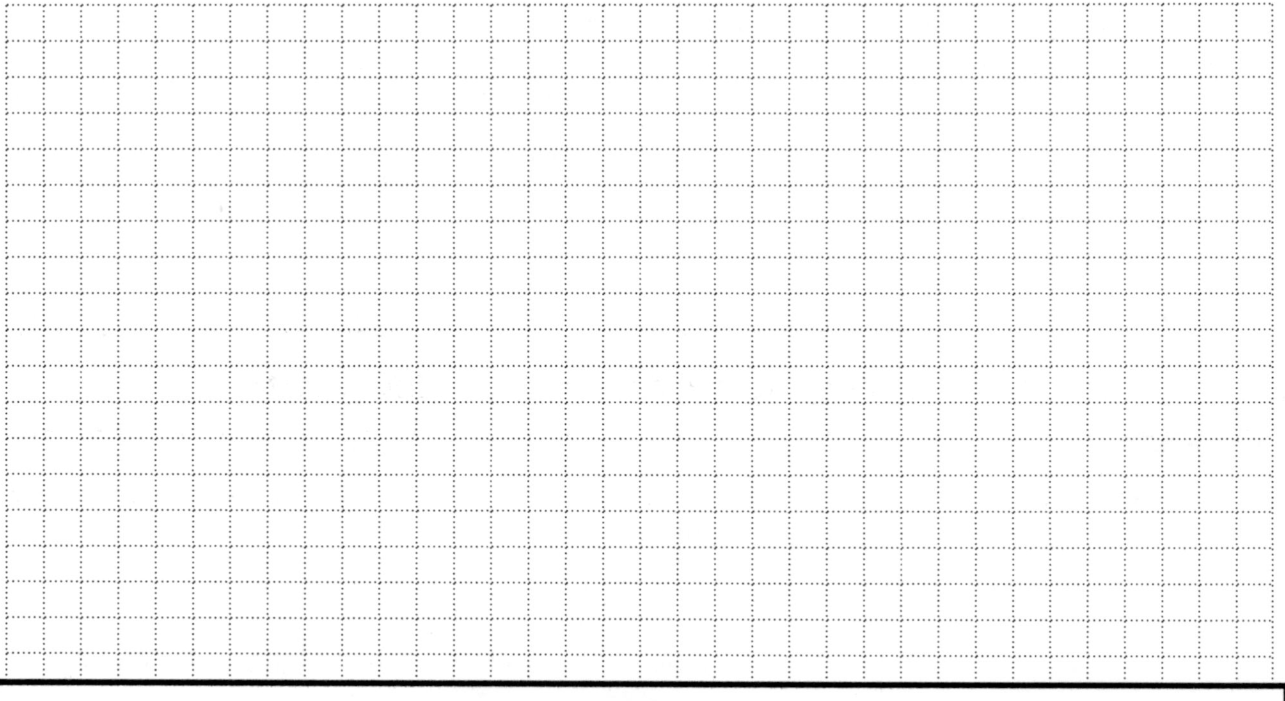

TABLE D					
roll to a distance of	time (s)				average speed (cm/s)
	1st trial	2nd trial	3rd trial	avg. time(s)	
25 cm					
50 cm					
Calculate the rate of accleration: a = (____ - ____) ÷ ____ = ________					
75 cm					
100 cm					
Calculate the rate of accleration: a = (____ - ____) ÷ ____ = ________					

To calculate the rate of acceleration use the formula . . . $\mathbf{a = (v_f - v_s) \div t}$ . . . where **a** is the rate of acceleration, $\mathbf{v_s}$ the average velocity calculated after rolling the ball to 25 cm, and $\mathbf{v_f}$ is the average velocity calculated after rolling the ball to 50 cm. **t** is the difference between the time the ball took to go 50 cm and the time it took to go 25 cm.

Name: ______________________________ **Period:**________ **Date:** ___/___/___

PS7 Review Quiz

Directions: Keep your eyes on your own work.
Read all directions and questions carefully.
THINK BEFORE YOU ANSWER!
Watch your spelling, be neat, and do the best you can.

CLASSWORK	(~40):	______
HOMEWORK	(~20):	______
CURRENT EVENT	(~10):	______
TEST	(~30):	______
TOTAL	(~100):	______

(A ≥ 90, B ≥ 80, C ≥ 70, D ≥ 60, F < 60)

LETTER GRADE: ______

TEACHER'S COMMENTS: ____

____

SPEED, VELOCITY, AND ACCELERATION

TRUE–FALSE FILL-IN: If the statement is true, write the word TRUE. If the statement is false, change the underlined word to make the statement true. *8 points*

______________ 1. Speed is called a <u>scalar</u> quantity.

______________ 2. Velocity is a <u>vector</u> quantity.

______________ 3. The velocity of a moving object includes information about the object's speed and <u>mass</u>.

______________ 4. <u>Speed</u> is any change in velocity.

______________ 5. <u>Acceleration</u> is distance divided by time.

______________ 6. The energy an object has <u>never</u> changes when its velocity changes.

______________ 7. A small object moving at high velocity <u>can</u> have more energy than a large object moving at low velocity.

______________ 8. The rate of an object's acceleration <u>can</u> stay the same while its velocity increases.

MATCHING: Choose the unit of measure on the right that matches the quantity on the left. *10 points*

_____ 9. length	(A)	grams per cubic centimeter
_____ 10. speed	(B)	grams
_____ 11. mass	(C)	meter
_____ 12. density	(D)	meters per second per second
_____ 13. acceleration	(E)	meters per second

PROBLEMS

Directions: Show all math calculations. Use correct units of measure. Put answers in the spaces provided!!

PROBLEM #1 —6 points

A car travels 500 kilometers in 10 hours. What is the average speed of the car during that time?

Final Answer: ____________________

PROBLEM #2— 6 points

A roller coaster is moving at 35 meters per second at the bottom of a hill. Five seconds later it nears the top of the hill moving at 5 meters per second. What is the deceleration of the roller coaster?

Final Answer: ____________________

______________________________ Student's Signature

______________________________ Parent's Signature

____/____/____ Date

PS8

SIR ISAAC NEWTON AND THE LAWS OF MOTION

TEACHER'S CLASSWORK AGENDA AND CONTENT NOTES

Classwork Agenda for the Week

1. Students will demonstrate that massive objects have inertia and show how the force of friction interferes with the motion of objects.
2. Students will show the relationship between force, mass, and acceleration.
3. Students will show how the force on an object, its mass, or acceleration can be calculated using Newton's Second Law of Motion.
4. Students will explain why the Law of Conservation of Momentum makes rocket flight possible.

Content Notes for Lecture and Discussion

Sir Isaac Newton (b. 1642; d. 1727) laid the foundation of **classical mechanics** in his **Principia** published in 1687. The volume contained a thorough description of the concepts of **mass, velocity, acceleration**, and **force** which Newton related to one another in his three **Laws of Motion**. In later publications Newton used those laws to derive his better known **Universal Law of Gravitation** which aided seventeenth and eighteenth century physicists in understanding the motion of the planets. Newton's **Law of Gravity**, however, was supplanted by the **Special and General Theories of Relativity** advanced by **Albert Einstein** (b. 1879; d. 1955) at the turn of this century. Einstein's Special and General Theories of Relativity gave a more complete description of how the universe works. Einstein explained that the observed motion of any body is dependent upon the motion of the observer. In short, motion is "relative." A discussion of Einstein's discoveries can be reserved for gifted and upper-level students, since Newton's Law of Gravity is sufficient to describe everyday observations where the motion of objects is well below that of the speed of light. Newton's Laws of Motion and Gravity was enough to put men on the moon. Newton is also well known for having invented the **infinitesimal calculus** which widened physicists' understanding of **light phenomena** and the **mechanics of planetary motion**. In the course of your class discussions throughout this unit, raise student awareness of the list of contributions made by one of the world's greatest mathematician-philosophers: Sir Isaac Newton.

Review with students the definition of **mass** as the amount of matter in an object. Tell them that all massive objects have a property called **inertia**. After a brief discussion of these terms perform the first two demonstrations outlined in Lesson #1 to show the inertial property of matter. Define inertia as the tendency of an object to remain at rest or in constant motion unless acted upon by an outside force. Explain that **friction**, the collision of atoms and molecules along opposing surfaces, is one force able to change the speed and direction of moving objects. Because all material surfaces are composed of atoms and molecules there is no such thing as a "perfectly smooth" surface. Discuss the motion of a space capsule travelling through space where there are no more than one or two atoms per cubic meter of interplanetary space. Outer space is a nearly perfect material vacuum; there is little matter to interfere with an object's motion through space. As a result, additional fuel is not needed to propel a spacecraft toward another planet once it has begun its journey after escaping earth's gravity. The craft does, however, require fuel to make changes in speed or direction, unless gravity is used to "steer" the craft. Explain that Galileo was the first to realize that an object in motion along a hypothetically "frictionless" surface would continue in a straight line at a constant velocity. His reasoning was as follows: (1) Objects rolling down an incline accelerate. (2) Objects rolling up an incline decelerate. Therefore, objects moving along a frictionless surface that is perfectly horizontal with the surface of the earth will neither accelerate nor decelerate but continue moving at a constant velocity. Use Lesson #1 to demonstrate how friction influences motion.

Lessons #2 and #3 include activities which help to illustrate Newton's **Second Law of Motion**. Begin by giving students a simple definition of **force** as a "push or pull" on an object. Then use the demonstrations in Lesson #2 to prompt discussion about how mass and velocity both influence the force an object applies when colliding with other objects. Use the formula **p = m × v** to calculate the **momentum** (**p**) of an object. In Lesson #3, show students how to use the formula **f = m × a** to calculate the force (**f**), mass (**m**), or acceleration (**a**) of an object when two of the variables are given. Explain that force is a "derived" unit of measure expressed in either "gram-centimeters per second squared" (i.e., g×cm/s^2) or "kilogram-meters per second squared" (i.e., kg×m/s^2). Since these units of measure are too cumbersome to use, introduce the term **dyne** to refer to 1 g×cm/s^2. One dyne is the force required to accelerate a 1-gram object at the rate of 1 centimeter per second every second. The term **newton** is used to describe 1 kg×m/s^2. One newton is the force required to accelerate a 1-kilogram object at the rate of 1 meter per second every second. One newton equals 100,000 dynes.

Lesson #4 includes several activities that demonstrate Newton's **Third Law of Motion**, a corollary of The Law of Conservation of Momentum. The Third Law can be stated in its most popular form as follows: For every action there is an equal and opposite reaction. Show students how to interpret problems like Problem #4 in the Homework and how to substitute given values for their appropriate variables in the following equation: $\mathbf{m_1v_1 + m_2v_2 = m_1v_3 + m_2v_4}$. Explain that rockets do not race toward the heavens by "pushing against the atmosphere" as one "pushes against the ground" while walking. The loss of mass from the rocket as gases are expelled from the vehicle cause the rocket to accelerate in the opposite direction in order to conserve the momentum of the whole system.

ANSWERS TO THE HOMEWORK PROBLEMS

1. p = m × v
   p = 5 kg × 25 m/s = 125 p (or momenta)
2. f = m × a
   f = 30 kg × 25m/s^2 = 750 kg × m/s^2 = 750 newtons
3. a = f ÷ m a = 225 newtons ÷ 15 kg = 15 m/s^2
4. Plug in the values found in the problem according to the following formula:

m_1v_1	+	m_2v_2	=	m_1v_3	+	m_2v_4
(20 kg × 15 m/s)	+	(10 kg × 5 m/s)	=	(20 kg × 5 m/s)	+	(10 kg × ?)
(300 p)	+	(50 p)	=	(100 p)	+	(10 kg × ?)

Since the momentum before and after the collision must be equal,

(350 p)	=	(100 p)	+	(10 kg × 25 m/s)

The velocity of the second ball after the collision, v_4, must be <u>25 m/s</u>.

ANSWERS TO THE END-OF-THE-WEEK REVIEW QUIZ

1. Newton
2. Newton
3. Newton
4. inertia
5. true
6. true
7. true
8. Third
9. 18 newtons
10. 32 dynes

PROBLEM

12 g mass	×	3 cm/s velocity	+	3 g mass	×	6 cm/s velocity	=	12 g mass	×	2 cm/s velocity	+	3 g mass	×	10 cm/s velocity
36 p			+	18 p			=	24 p			+	(must be 30 p)		
		because 54 p must					=	54 p						

Final Answer: <u>10 cm/s</u>

Name: ______________________________ Period:________ Date: ___/___/___

PS8 FACT SHEET

SIR ISAAC NEWTON AND THE LAWS OF MOTION

CLASSWORK AGENDA FOR THE WEEK

(1) Demonstrate that massive objects have inertia and show how the force of friction interferes with the motion of objects.
(2) Show the relationship between force, mass, and acceleration.
(3) Show how the force on an object, its mass, or acceleration can be calculated using Newton's Second Law of Motion.
(4) Explain why the Law of Conservation of Momentum makes rocket flight possible.

The work of the great seventeenth century mathematician-philosopher, Sir Isaac Newton (born, 1642; died, 1727), laid the foundation of **classical physics**. Newton summarized the motion of objects in three **Laws of Motion**. His **Universal Law of Gravitation** is still used today to plan missions to the moon and the planets of our solar system. Newton's physical description of the universe is called **classical mechanics**.

Newton's **First Law of Motion** is called the **Law of Inertia**. **Inertia** is the tendency of matter to remain at rest or in constant motion unless acted upon by some outside force. The reason for this is that all moving objects have **momentum**. Momentum (symbolized by the letter **p**) is the product of an object's mass **(m)** and velocity **(v)**:

$$\mathbf{p = m \times v}$$

A **force** is a push or pull on an object. Newton observed that a force **(f)** is needed to change the position, velocity, or direction of a moving object. He summarized this observation with his **Second Law of Motion**. The Second Law recognizes that the force on an object is "proportional" to the object's mass **(m)** and acceleration **(a)**:

$$\mathbf{f = m \times a}$$

The more force that is applied to an object, the more it accelerates. The more massive an object is, the larger the force needed to accelerate it.

Newton's **Third Law of Motion** follows from the **Law of Conservation of Momentum**. This law predicts that the total "momentum" of many individual objects in a "closed system" remains constant no matter how the objects interact. The total momentum of a group of objects does not change. This law can be symbolized by the following mathematical equation:

$$\mathbf{m_1v_1 + m_2v_2 = m_1v_3 + m_2v_4}$$

In the equation above, "m_1v_1" is the momentum of object "m_1" before a collision. And, "m_2v_2" is the momentum of object "m_2" before a collision. After a collision of the two objects, their velocities change. But the total momentum of the objects—"m_1v_3" plus "m_2v_4"—will be the same as the total momentum of the objects before they collided—"m_1v_1" plus "m_2v_2".

The science of rocketry takes advantage of Newton's Third Law of Motion. Gases expelled from the exhaust nozzle of a rocket booster cause the rocket to lose mass and lift off the launch pad. In order "to conserve" the momentum of the system, the rocket increases its velocity and moves in the opposite direction as the expelled gases. Newton's Third Law is sometimes stated as follows: For every action there is an equal and opposite reaction.

Homework Directions

SHOW ALL MATHEMATICAL FORMULAS AND CALCULATIONS IN SOLVING PROBLEMS #1, #2, #3 AND #4. BE SURE TO INCLUDE CORRECT UNITS OF MEASURE WITH YOUR ANSWER.

1. What is the momentum of a 5-kilogram object moving at a velocity of 25 meters per second?
2. What is the force on a 30-kilogram object accelerating at the rate of 25 meters per second per second?
3. Find the rate of acceleration of a 15-kilogram mass subjected to a force of 225 newtons.
4. A solid ball with a mass of 20 kilograms moving at a velocity of 15 meters per second collides with another solid ball with a mass of 10 kilograms moving at 5 meters per second. If the first ball moves away from the collision with a velocity of 5 meters per second, how fast will the second ball speed away from the collision? NOTE: Assume that the masses of the two objects in this problem are not changed by the collision.

Assignment due: __________

______________________ Student's Signature

______________________ Parent's Signature

____/____/____ Date

PS8 Lesson #1

SIR ISAAC NEWTON AND THE LAWS OF MOTION

Work Date: ____/____/____

LESSON OBJECTIVE

Students will demonstrate that massive objects have inertia and show how the force of friction interferes with the motion of objects.

Classroom Activities

On Your Mark!

Introduce students to the concept of inertia by setting a penny on top of a playing card placed on top of a glass or beaker. Quickly knock the playing card out from under the coin with the flick of an index finger, allowing the coin to drop vertically down into the glass. Begin a discussion that helps students to analyze the forces involved in this simple demonstration. What forces kept the coin at rest on the playing card? Friction? Gravity? Both? Why didn't the coin fly away with the card? Did the coin's own "stubbornness" prevent it from doing so? Or, does the coin have some intrinsic property that favors its remaining at rest? Next, place the beaker on a paper towel and quickly snap the paper towel out from under it like a magician snapping a table cloth away from a table set with fine linen, crystal, and china. Is the magician really performing "magic" or just demonstrating Newton's First Law of Motion: The Law of Inertia? Have students copy The Law of Inertia onto Journal Sheet #1: A body will tend to remain at rest or in constant motion unless acted upon by some outside force.

Get Set!

Continue the discussion by reviewing the concept of mass and add that all material objects have the intrinsic property called inertia: the tendency to resist a change in motion. Explain that all objects are really in motion relative to other objects even though they might appear to be at rest. Ask students: "Are we at rest in this classroom?" Answer: "No. Because the earth is rotating as it moves through space around the sun. The sun, also, is moving through the universe." Explain that friction, caused by the collision of atoms along the surfaces of objects, is a major force that impedes movement.

Go!

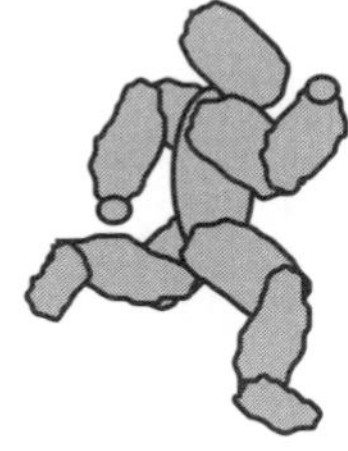

Show students how to set up the experiment pictured in Figure A of Journal Sheet #1. Challenge them to drag a beaker filled with different amounts of liquid across several different surfaces (i.e., a slick linoleum floor, a piece of old carpet, a stretch of sandpaper, etc.). Have them record in Table A the amount of force registered by the spring scale at the instant each beaker overcomes friction with the underlying surface and begins to move. Students will discover that more force is required to move heavier objects across rougher surfaces. Have plenty of towels handy to clean up the watery mess!

Materials

pennies, playing cards, balances, beakers, spring scales, string, sandpaper, old pieces of carpet

Name: ______________________________ Period:________ Date: ____/____/____

PS8 Journal Sheet #1

SIR ISAAC NEWTON AND THE LAWS OF MOTION

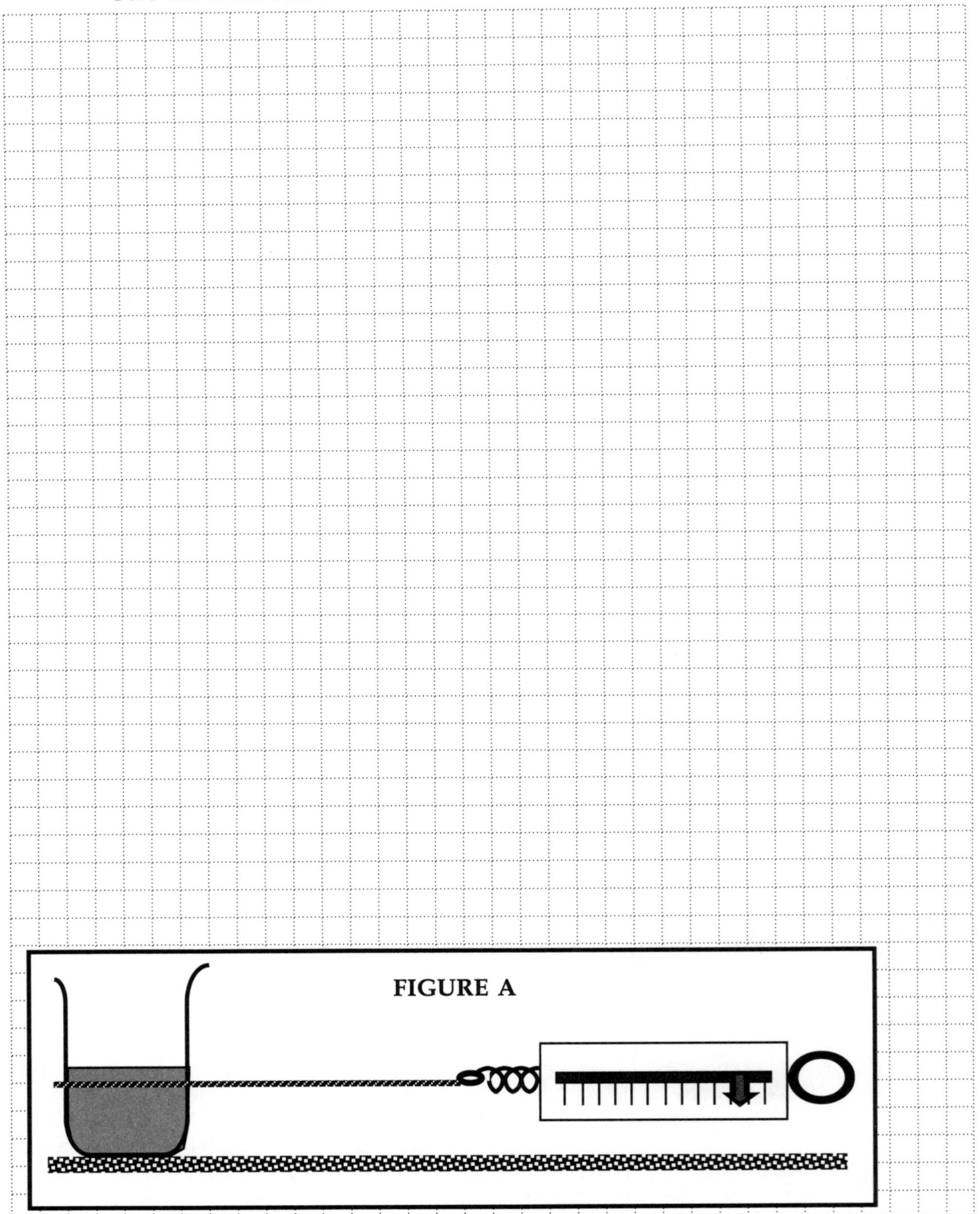

SIR ISAAC NEWTON AND THE LAWS OF MOTION

Work Date: ___/___/___

LESSON OBJECTIVE

Students will show the relationship between force, mass, and acceleration.

Classroom Activities

On Your Mark!

Begin a discussion of momentum by asking the following question: "Which object is likely to do more damage to a brick wall: (1) a 5-kilogram bowling ball moving at 1 meter per second, or (2) a 0.010-kilogram bullet moving at 1,000 meters per second. Student answers will vary. In fact, the momentum of the bullet, and therefore the force it applies to the wall, is greater than that of the bowling ball. Write the formula for momentum, $\mathbf{p}=\mathbf{m} \times \mathbf{v}$, on the board and explain the meaning of each variable. Substitute the mass and velocity values for the bowling ball and bullet into the equations and show students why the bullet would do more damage. Upon instant deceleration as it hits the wall, the force of the bowling ball is 5 newtons. The force of the bullet is 10 newtons.

Get Set!

Show students how to set up the experiment shown in Figure B on Journal Sheet #2. Demonstrate how to propel the "shot coin" at a larger or smaller "target coin" at a low, medium, or high rate of speed. Have them record the distance travelled by the target coin at each trial on Journal Sheet #2.

Go!

Have students perform the experiment, making sure that they "shoot" coins of different sizes with approximately the same degree of force in each of the three force categories (i.e., low, medium, and high). Although this experiment is more of a "qualitative" demonstration than a "quantitative" one, students should realize that (1) more massive objects require more force to move them a greater distance, and that (2) the force of an object increases with its velocity. Summarize their results on the board by writing Newton's Second Law of Motion: $\mathbf{f} = \mathbf{m} \times \mathbf{a}$. Explain that the force required to accelerate an object is directly proportional to its mass and the degree to which you wish to accelerate it.

Materials

metric rulers, coins of different sizes

Name: ______________________________ Period:________ Date: ___/___/___

PS8 Journal Sheet #2

SIR ISAAC NEWTON AND THE LAWS OF MOTION

TABLE A			
shot coin	speed (low, med, high)	target coin	distance

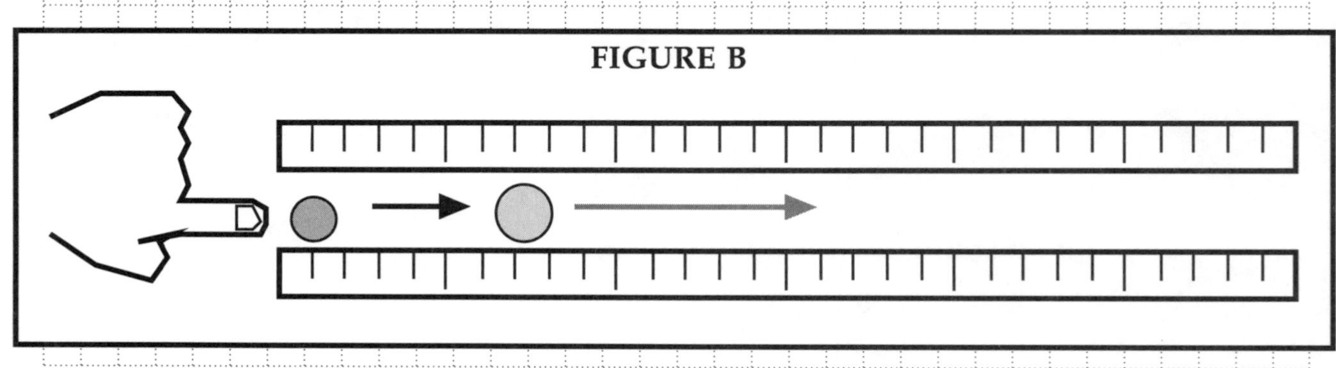

PS8 Lesson #3

SIR ISAAC NEWTON AND THE LAWS OF MOTION

Work Date: ____/____/____

LESSON OBJECTIVE

Students will show how the force on an object, its mass, or acceleration can be calculated using Newton's Second Law of Motion.

Classroom Activities

On Your Mark!

Begin with a review of the experiment recorded on Journal Sheet #2. Review the meanings of **inertia** and **momentum** by asking students to share an anecdote in which their inertia or momentum (i.e., on a skateboard or bicycle) became thrillingly apparent. Share a personal experience of your own to animate the discussion (i.e., a minor car accident or sports experience). Write Newton's Second Law of Motion on the board: **f = m** × **a**. Review the meaning of each variable.

Get Set!

Have students copy your board notes on Journal Sheet #3 along with your explanation of each variable's correct unit of measure.

concepts:	**force**	= **mass** × **acceleration**
new unit of measure:	**dyne**	= **gram** × **centimeter/second squared**
new unit of measure:	**newton**	= **kilogram** × **meter/second squared**

Explain that **force**, defined as a "push or pull" on an object, is measured in **dynes** (CGS) or **newtons** (MKS). Work through a few problems to help students calculate the amount of force required to accelerate an object of a particular mass. Use the problems on Journal Sheet #3 and create other examples.

Go!

Have students work cooperatively to solve the problems in the table on Journal Sheet #3. Help them to correct their mistakes as you circulate around the room. Emphasize the use of the correct units of measure!

Materials

Journal Sheet #3

ANSWERS TO PROBLEMS ON JOURNAL SHEET #3

	FORMULA	UNDERLINED ANSWERS AND CALCULATIONS				
1.	f = m × a	<u>0 newtons</u>	=	1,000 kg	×	0 m/s^2 (the object is not accelerating!)
2.	p = m × v	<u>250,000 momenta</u>	=	1,000 kg	×	250 m/s^2
3.	f = m × a	<u>52.5 newtons</u>	=	15 kg	×	3.5 m/s^2
4.	f = m × a	80 newtons	=	<u>40 kg</u>	×	2 m/s^2
5.	f = m × a	100 dynes	=	4 g	×	<u>25 cm/s^2</u>
6.	f = m × a	<u>300,000 dynes</u>	=	5 g	×	60,000 cm/s^2 (600 m × 100 cm/m = 60,000 cm)
7.	f = m × a	<u>490 newtons</u>	=	50 kg	×	9.8 m/s^2 (*Note:* 9.8 m/s^2 is the acceleration due to earth's gravity at sea level)

Name: ______________________ Period: ________ Date: ___/___/___

PS8 Journal Sheet #3

SIR ISAAC NEWTON AND THE LAWS OF MOTION

PROBLEMS

Directions: Work out each problem in the open space provided on this Journal Sheet. Show all formulas and circle the final answer to each problem.

1. How much force is being applied to a 1,000-kilogram meteor moving through interstellar space at a constant velocity of 250 meters per second?
2. What is the momentum of the meteor in Problem #1?
3. How much force is being applied to a 15-kilogram bowling ball rolling down a hill at 3.5 meters per second per second?
4. Find the mass of an object accelerating at 2 meters per second per second if the force being applied to the object is 80 newtons.
5. Find the rate of acceleration of a 4-gram mass being pulled by a 100-dyne force.
6. What is the force on a 5-gram bullet decelerating at 600 meters per second per second?
7. What is the force applied to a falling 50-kilogram mass falling at 9.8 meters per second per second?

PS8 Lesson #4

SIR ISAAC NEWTON AND THE LAWS OF MOTION

Work Date: ____/____/____

LESSON OBJECTIVE

Students will explain why the Law of Conservation of Momentum makes rocket flight possible.

Classroom Activities

On Your Mark!

Begin discussion with a review of the concept of momentum: **p = m x v**. Explain that the **Law of Conservation of Momentum** is one of the foundations of the science of physics. This Law can be stated as follows: The total momentum of all the moving objects in a closed system never changes. It is constant. In other words, if one were to add up all of the calculated momenta of the objects in the universe (which is, of course, not possible) the result would be the same regardless of how the motion of individual objects has changed over time due to collisions. It follows from this law, therefore, that "for every action there must be an equal and opposite reaction" that cancels it out. The latter quote is the more popular description of this law. Ask students this question: What would happen to a row boat on a lake if the oarsman in the boat suddenly decided to leap into the water? Would the boat move? If so, in which direction?

Get Set!

Write the following formula on the board and have students record it in Journal Sheet #4:

$$\mathbf{m_1v_1 + m_2v_2 = m_1v_3 + m_2v_4}$$

Explain the meaning of each variable in the equation: $\mathbf{m_1}$ = the mass of one object; $\mathbf{v_1}$ = the velocity of that object before a collision; $\mathbf{m_2}$ = the mass of a second object; $\mathbf{v_2}$ = the velocity of the second object before a collision; $\mathbf{v_3}$ = the velocity of the first object after a collision with m_2; and, $\mathbf{v_4}$ = the velocity of the second object after a collision with the first object. Create a problem like the Homework Problem to assist students in recognizing the meaning of each variable and how to solve for one of the variables when all of the others are known.

Go!

Have students construct the "rocket balloons" illustrated in Figure C on Journal Sheet #4 and conduct a contest to see whose balloon can fly the farthest across the room.

Materials

string, balloons, paper or plastic drinking straws, tape, chairs

Name: ______________________ **Period:**________ **Date:** ___/___/___

PS8 JOURNAL SHEET #4

SIR ISAAC NEWTON AND THE LAWS OF MOTION

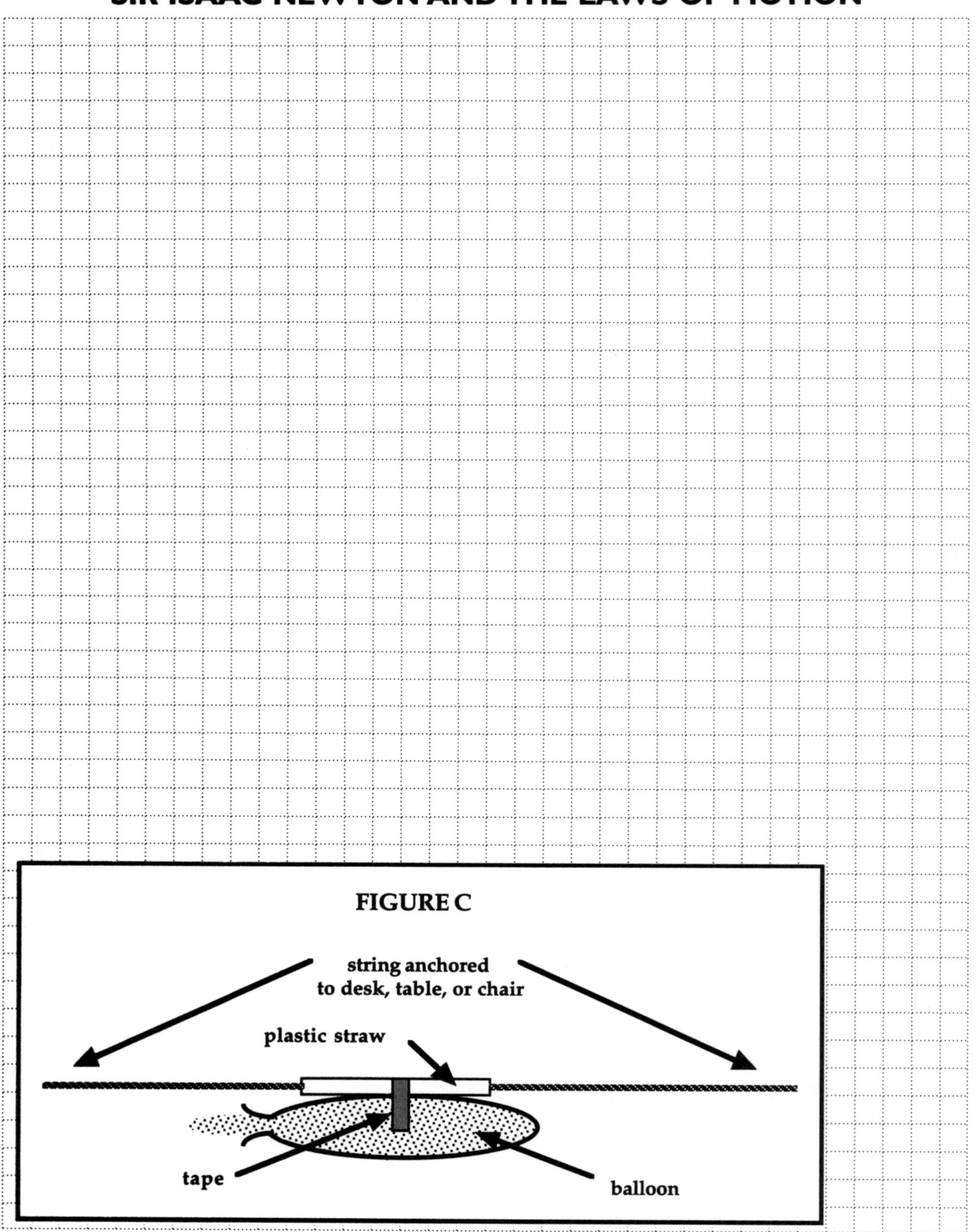

Name: ______________________________ **Period:**________ **Date:** ____/____/____

PS8 Review Quiz

Directions: Keep your eyes on your own work.
Read all directions and questions carefully.
THINK BEFORE YOU ANSWER!
Watch your spelling, be neat, and do the best you can.

CLASSWORK	(~40):	_______
HOMEWORK	(~20):	_______
CURRENT EVENT	(~10):	_______
TEST	(~30):	_______
TOTAL	(~100):	_______

(A ≥ 90, B ≥ 80, C ≥ 70, D ≥ 60, F < 60)

LETTER GRADE: _______

TEACHER'S COMMENTS: ____

____

SIR ISAAC NEWTON AND THE LAWS OF MOTION

TRUE–FALSE FILL-IN: If the statement is true, write the word TRUE. If the statement is false, change the underlined word to make the statement true. *20 points*

______________ 1. <u>Galileo</u> showed that force, mass, and acceleration are related.

______________ 2. <u>Einstein</u> summarized the motion of objects in three Laws of Motion.

______________ 3. <u>Copernicus'</u> Universal Law of Gravitation explains how the planets stay in orbit around the sun.

______________ 4. All objects have <u>momentum</u> which is the tendency of matter to stay at rest or in constant motion until acted upon by some outside force.

______________ 5. The <u>momentum</u> of an object is the product of its mass and velocity.

______________ 6. The total momentum of a group of objects in a closed system <u>does not</u> change.

______________ 7. The <u>force</u> on an object is the product of its mass and acceleration.

______________ 8. Newton's <u>First</u> Law of Motion is sometimes stated as follows: For every action there is an equal and opposite reaction.

______________ 9. A 6-kilogram object accelerating at the rate of 3 meters per second per second has a force of <u>9 dynes</u>.

______________ 10. An 8-gram object accelerating at the rate of 4 centimeters per second per second has a force of <u>2 newtons</u>.

PROBLEM

Directions: Fill in the blanks below to help solve the following problem. Show all math calculations in the margins and put your final answer in the blank provided near the bottom of the page. Remember to give a correct unit of measure with your final answer! *10 points*

A solid ball with a mass of 12 grams is moving at a velocity of 3 centimeters per second. It collides with another solid ball having a mass of 3 grams moving at 6 centimeters per second. The first ball moves away from the collision with a velocity of 2 centimeters per second. How fast will the second ball speed away from the collision? NOTE: Assume that the masses of the two objects are not changed by the collision.

_____ × _____ + _____ × _____ = _____ × _____ + _____ × _____
mass   velocity   mass   velocity   mass   velocity   mass   velocity

Final Answer: ____________

____________ Student's Signature

____________ Parent's Signature

___/___/___ Date

PS9

GALILEO, NEWTON, AND THE LAW OF GRAVITY

TEACHER'S CLASSWORK AGENDA AND CONTENT NOTES

Classwork Agenda for the Week

1. Students will demonstrate how to find the center of gravity of an oddly shaped object.
2. Students will show that objects accelerate at a constant rate toward the earth.
3. Students will show that objects accelerate toward the surface of the earth at a constant rate regardless of their masses.
4. Students will find the force of gravity between two massive objects using Sir Isaac Newton's formula for calculating gravitational force.

Content Notes for Lecture and Discussion

The Greek philosopher **Aristotle** (b. 384 B.C.; d. 322 B.C.) laid down a set of "natural laws" that were accepted as fact for well over 2,000 years. Although Aristotle's arguments were impeccably logical and thorough, his explanations of many natural phenomena turned out, for the most part, to be wrong! Aristotle believed that knowledge was best acquired by careful study with the senses and introspective examination by the mind. He never bothered to "test" his ideas through experiment but accepted that which could be logically conceived as the truth. In the seventeenth century, Aristotle's "science" was completely overhauled by the rigorous analysis of one of the world's first great "experimental" scientists: Galileo Galilei.

Galileo Galilei (b. 1564; d. 1642) stunned his contemporaries by demonstrating the falsity of Aristotle's beliefs. Among them was the idea that objects of differing mass fell to earth at different rates. According to Aristotle, a ball ten times heavier than another would fall to earth ten times faster; but he never bothered to experiment. Galileo did. With his Aristotelian mentors looking on, Galileo dropped balls of differing mass from the Leaning Tower of Pisa in Italy. They hit the ground at almost exactly the same time—the slight difference in their drop times was due to the effects of air friction. Much of Galileo's work was done using ramps to measure the acceleration of "falling" objects, since the rate of acceleration of a free falling body was much too difficult to measure without an adequate timing device. Galileo timed the descent of balls rolling down a ramp by singing. He marked the position of the rolling ball on every beat of his song and determined the increase in distance covered with each passing unit of time to be the same. Thus, the rate of acceleration due to gravity was constant, regardless of the mass of the falling body. It was Galileo's work that laid the foundation for Sir Isaac Newton's ingenious formulation of his **Universal Law of Gravitation**.

Newton reasoned that individual objects must be responsible for the resultant force between them. How could the earth "know," for example, to pull a ten-pound object down with greater force than a one-pound object so that both would reach the ground at the same time? Use the analogy introduced in Lesson #1 to explain Newton's reasoning to students. Newton's Universal Law of Gravitation was finally formulated as follows:

$$\mathbf{F_g = G(m_1 \times m_2) \div r^2}$$

where $\mathbf{F_g}$ is the force of gravitational attraction between two objects, $\mathbf{m_1}$ is the mass of one object, $\mathbf{m_2}$ is the mass of another object, and **r** is the distance between them. In this formula, the masses of both objects determine the force of gravity between them. In addition, an increase in either mass will increase the force of gravity between the objects. The force of gravity is "directly proportional" to the masses of the objects. Inversely, an increase in the distance between the objects results in a decrease of the force between the objects. **G** is the **gravitational constant** expressed in newton–meter^2/kg^2. The gravitational constant is the force or gravity between two 1-kilogram objects placed exactly 1 meter apart. In 1798, Henry Cavendish (b. 1731; d. 1810) determined that force to be equal to 6.67×10^{-11}newtons—an extremely small amount of force! This simple formulation of gravitational force between any two massive objects still serves to describe the motion of spacecraft, satellites, and planetary bodies. At the start of the twentieth century, however, **Albert Einstein** (b. 1879; d. 1955) revolutionized our thinking about the universe with his **General Theory of Relativity**. Einstein's notion supplanted Newton's Law of Gravity.

According to Einstein, gravity is a "curvature" in the space-time continuum resulting from the presence of massive bodies. Objects accelerate toward one another as a result of the "warping" of space between them. Modern physicists still search for the elusive "graviton," the force particle which may mediate the force of gravity between massive objects.

Since objects near the surface of the earth are much smaller in comparison to the size of the earth (which has a mass of about 6×10^{24} kilograms), the mass of small bodies can be neglected when accounting for the force of gravity being exerted upon them. **Weight** is therefore defined as the force of gravity on an object near the earth's surface and can be adequately calculated using the following simple formula:

$$\mathbf{W = m \times g}$$

where **W** is the weight of the object in dynes or newtons, **m** is the mass of the object in grams or kilograms, and **g** is the acceleration of gravity near the earth's surface (**9.8 m/s^2** or **980 cm/s^2**).

ANSWERS TO THE HOMEWORK PROBLEMS

1. $F_g = G(m_1 \times m_2) \div r^2$
   mass of the earth= 6.0×10^{24} kilograms
   mass of the moon = 7.0×10^{22} kilograms
   distance between the earth and moon = 3.84×10^5 kilometers = 3.84×10^8 meters
   $F_g = (6.67 \times 10^{-11})(6.0 \times 10^{24})(7.0 \times 10^{22}) \div (3.84 \times 10^8)(3.84 \times 10^8)$ or
   $F_g \approx (7 \times 10^{-11})(6.0. \times 10^{24})(7.0 \times 10^{22}) \div (4 \times 10^8)(4 \times 10^8) \approx 2 \times 10^{20}$ newtons
2. W = m × g
   m = 30 kg
   g = 9.8 m/s_2
   W = 294 newtons
3. Answers will vary but should include a description of the effect of air friction on the parachute's descent.

ANSWERS TO THE END-OF-THE-WEEK REVIEW QUIZ

1. true
2. true
3. true
4. true
5. true
6. weight
7. are not
8. 500 newtons
9. true
10. true
11. B
12. A
13. C
14. D
15. B

Name: ______________________________ Period:________ Date: ____/____/____

PS9 Fact Sheet

GALILEO, NEWTON, AND THE LAW OF GRAVITY

CLASSWORK AGENDA FOR THE WEEK

(1) Find the center of gravity of an oddly shaped object.
(2) Show that objects accelerate toward the surface of the earth at a constant rate.
(3) Show that objects accelerate toward the surface of the earth at a constant rate regardless of their masses.
(4) Find the force of gravity between two massive objects using Sir Isaac Newton's formula for calculating gravitational force.

The Italian scientist, **Galileo Galilei** (born, 1564; died, 1642), dropped objects off the Leaning Tower of Pisa to test the laws of nature. He observed that objects of different mass, dropped from the same height at the same time, hit the ground at the same instant. Less than 100 years later, Sir Isaac Newton (born, 1642; died, 1727), used Galileo's results to explain how gravity works.

Newton explained that the force of gravity between two objects, like the earth and you, does not depend on the mass of the earth alone. Both the mass of the earth and your mass determine the force of gravity between the two of you. You are "pulling up" on the earth at the same time it is "pulling down" on you! Newton's **Universal Law of Gravitation** can be summarized by the following mathematical formula:

$$\mathbf{F_g = G(m_1 \times m_2) \div r^2}$$

where $\mathbf{F_g}$ is the force of gravitational attraction between two objects, $\mathbf{m_1}$ is the mass of one object, $\mathbf{m_2}$ is the mass of the other object, and **r** is the distance between them. **G** is the **gravitational constant**. The gravitational constant is the force or gravity between two 1-kilogram objects placed exactly 1 meter apart. That force is equal to 6.67×10^{-11} newtons—an extremely small amount of force! The force of gravity between any two objects in the universe, from the tiniest bacteria to the largest star, can be calculated using Newton's formula.

Weight is a measure of the force of gravity on an object near the earth's surface. The amount of matter in an object is called **mass**. The weight of an object near the earth's surface is the force of the earth's gravitational pull on that object. Since the earth is much larger than objects near its surface, the force of that pull can be expressed simply as follows:

$$\mathbf{W = m \times g}$$

where **W** is the weight of the object in dynes or newtons, **m** is the mass of the object in grams or kilograms, and **g** is the acceleration of gravity near the earth's surface. The acceleration of gravity near the earth's surface is equal to 9.8 m/s^2 or 980 cm/s^2.

Homework Directions

SHOW ALL MATHEMATICAL FORMULAS AND CALCULATIONS IN SOLVING PROBLEMS #1 and #2. BE SURE TO INCLUDE CORRECT UNITS OF MEASURE WITH YOUR ANSWER.

1. The mass of the earth is approximately 6.0×10^{24} kilograms. The mass of the moon is approximately 7.0×10^{22} kilograms. The distance between the earth and moon is approximately 3.84×10^{5} kilometers. What is the force of gravity between the earth and the moon?

   Assignment due: __________

2. What is the weight in newtons of a 30-kilogram object near the surface of the earth?

   Assignment due: __________

3. A French physicist, L. Sébastien Lenormand made the first successful parachute jump off a tower in 1783. Write a paragraph of no less than 50 words that explains why the scientist did not plummet to earth at a constant acceleration due to gravity.

   Assignment due: __________

______________________	______________________	___/___/___
Student's Signature	Parent's Signature	Date

PS9 Lesson #1

GALILEO, NEWTON, AND THE LAW OF GRAVITY

Work Date: ____/____/____

LESSON OBJECTIVE

Students will demonstrate how to find the center of gravity of an oddly shaped object.

Classroom Activities

On Your Mark!

Open a discussion of the advantages and disadvantages of living in a universe that has gravity (i.e., we don't float away into outer space, yet we can trip and fall on our head). What determines the strength of gravity? If the force of gravity is greater as the "size" or "mass" of the object gets bigger, where exactly does the force originate?

Get Set!

Demonstrate how to find an object's "center of gravity" by referring to Figure A in Journal Sheet #1.

Go!

Have students perform the activity described in Figure A on Journal Sheet #1.

Materials

oddly cut pieces of cardboard, string, metal nuts, ring stands and clamps, knitting or dissecting needles

Name: ______________________________ Period:________ Date: ___/___/___

PS9 Journal Sheet #1

GALILEO, NEWTON, AND THE LAW OF GRAVITY

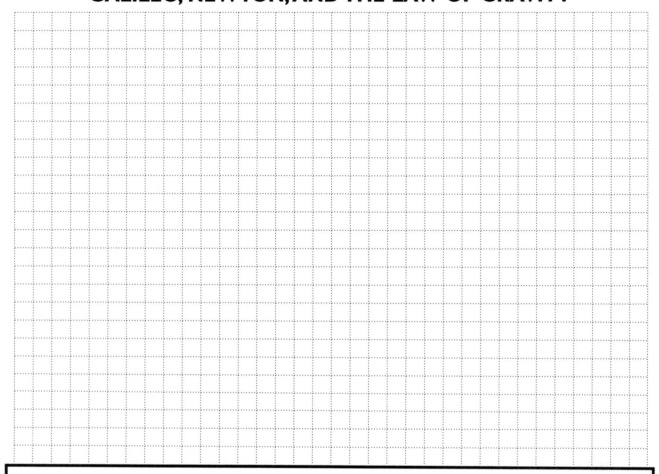

FIGURE A

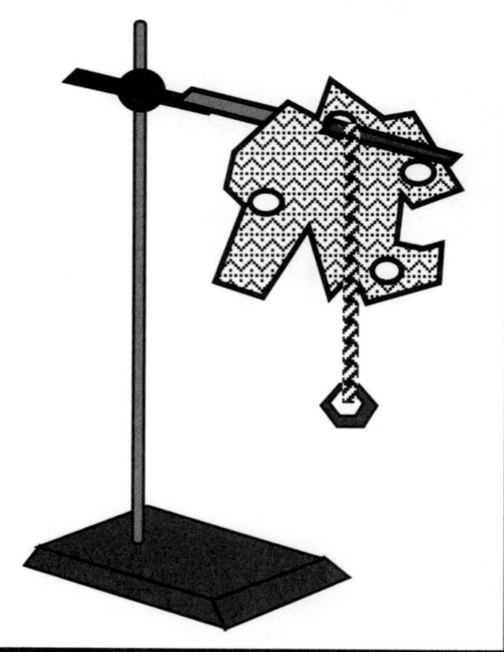

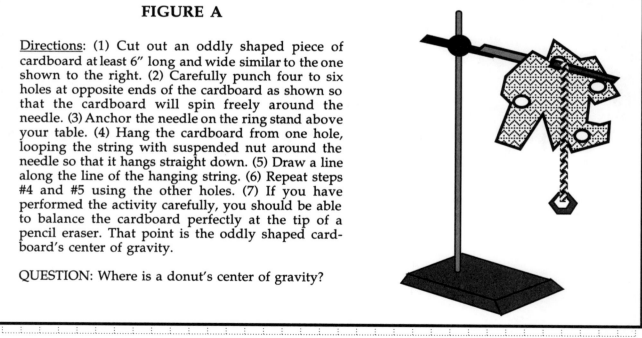

Directions: (1) Cut out an oddly shaped piece of cardboard at least 6" long and wide similar to the one shown to the right. (2) Carefully punch four to six holes at opposite ends of the cardboard as shown so that the cardboard will spin freely around the needle. (3) Anchor the needle on the ring stand above your table. (4) Hang the cardboard from one hole, looping the string with suspended nut around the needle so that it hangs straight down. (5) Draw a line along the line of the hanging string. (6) Repeat steps #4 and #5 using the other holes. (7) If you have performed the activity carefully, you should be able to balance the cardboard perfectly at the tip of a pencil eraser. That point is the oddly shaped cardboard's center of gravity.

QUESTION: Where is a donut's center of gravity?

GALILEO, NEWTON, AND THE LAW OF GRAVITY

Work Date: ____/____/____

LESSON OBJECTIVE

Students will show that objects accelerate at a constant rate toward the earth.

Classroom Activities

On Your Mark!

Describe the "common sense laws" deduced by Aristotle and ask the class "Which object would fall faster: a bowling ball or baseball?" Discuss the value of experimentation as a means of clarifying ideas about the way nature behaves.

Get Set!

Perform the following demonstration: (1) Put a golf ball and a ping pong ball (or other differentially weighted objects of similar size) on a small tray. (2) Holding the tray in one hand, stand on a ladder, solid table, or chair (taking care to follow your district's safety regulations). (3) Instruct students to listen carefully for the number of clicks as the objects hit the ground. (4) Quickly pull aside the tray to release the balls at the same instant. With practice, you should be able to show that both objects hit the ground at the same time (i.e., one click is heard by all). Following discussion of the demonstration, describe Galileo's Leaning Tower of Pisa experiments. Prepare students to perform the lesson activity shown on Journal Sheet #2.

Go!

Have students perform the activity described in Figure B on Journal Sheet #2. Have them record their results in Table A. Refer to the Sample below. How much farther did the object roll at each time interval compared to the distance it rolled in the previous time interval? Answer: 2 centimeters. Therefore, the acceleration of the ball due to gravity at this angle of incline is **2 cm/s/s**. The rate of acceleration is constant.

SAMPLE TABLE A

full swing	distance rolled (to nearest cm)			average distance rolled in this time interval	distance at this time interval minus distance of last time interval
	1st trial	2nd trial	3rd trial		
1	5	6	5	5	5
2	12	13	12	12	7
3	20	21	21	21	9
4	33	32	32	32	11

Compare the numbers in this column. →

Materials

marbles, tray, metric rulers, masking tape, string, metal nut

Name: ______________________ Period:________ Date: ___/___/___

PS9 Journal Sheet #2

GALILEO, NEWTON, AND THE LAW OF GRAVITY

low incline						TABLE A	higher incline					
full swings	distance rolled (to nearest cm) 1st trial	2nd trial	3rd trial	average distance rolled in this time interval	distance at this time interval minus distance of last time interval		full swings	distance rolled (to nearest cm) 1st trial	2nd trial	3rd trial	average distance rolled in this time interval	distance at this time interval minus distance of last time interval

Compare the numbers in this column → Compare the numbers in this column →

How much farther did the object roll at each time interval compared to the distance it rolled in the previous time interval?

low incline:_______ higher incline:_______

(Your answer is the acceleration of the ball at this angle of incline.)

FIGURE B

[-- 15 cm --]

Directions: (1) Tape together two or three metric rulers to form a "sluice" for your rolling marble. Cover the edges of the rulers with masking tape so that your marking pen will not discolor the ruler when you take measurements. (2) Prop the rulers up on a thin book. The more shallow the incline, the easier it will be to measure your rate of acceleration in the same way Galileo did. (3) Practice with your "pendulum clock" to make sure you and your partner can release and stop both ball and timer nut at the same time (i.e., "Ready … set … drop …stop!"). (4) On the masking tape, mark the distance rolled by the marble at one, two, three, and four full swings (i.e., back and forth) of the timer. Take three measures of distance for each full swing of the timer. (5) Raise the height of the incline slightly and repeat steps #1 through #4.

PS9 Lesson #3

GALILEO, NEWTON, AND THE LAW OF GRAVITY

Work Date: ____/____/____

LESSON OBJECTIVE

Students will show that objects accelerate toward the surface of the earth at a constant rate regardless of their masses.

Classroom Activities

On Your Mark!

Review the results of the activity performed in Lesson #2. If students performed the activity carefully, they should have discovered that—with few errors—their marbles increased the distance travelled with every "tick of their clock" by the same interval. Refer to the Sample Table A in Lesson #2. Refer to the demonstration performed at the start of Lesson #2 in which you dropped the ping pong ball and golf ball.

Get Set!

Review the procedure used in the activity performed in Lesson #2 and explain that different masses will be used this time to see if Galileo's experiments can be "replicated." After the activity explain that objects near the surface of the earth are much smaller in comparison to the size of the earth (which has a mass of about 6×10^{24} kilograms). Therefore, their masses can be neglected when accounting for the force of gravity being exerted upon them. **Weight** is therefore defined as the force of gravity on an object near the earth's surface and can be adequately calculated using the following simple formula, $\mathbf{W = m \times g}$, where **W** is the weight of the object in dynes or newtons, **m** is the mass of the object in grams or kilograms, and g is the acceleration of gravity near the earth's surface **($9.8\ m/s^2$ or $980\ cm/s^2$)**.

Go!

Have students repeat the activity performed in Lesson #2 using a golf ball and ping pong ball, marbles, or steel ball bearings of differential mass. Have them record their results in Table B on Journal Sheet #3. They will discover that the acceleration due to gravity at a particular angle of incline is the same regardless of the masses of the objects.

Materials

marbles, golf balls, ping pong balls, or steel ball bearing of differential masses, metric rulers, masking tape, string, metal nut

Name: ______________________________ Period:________ Date: ____/____/____

PS9 Journal Sheet #3

GALILEO, NEWTON, AND THE LAW OF GRAVITY

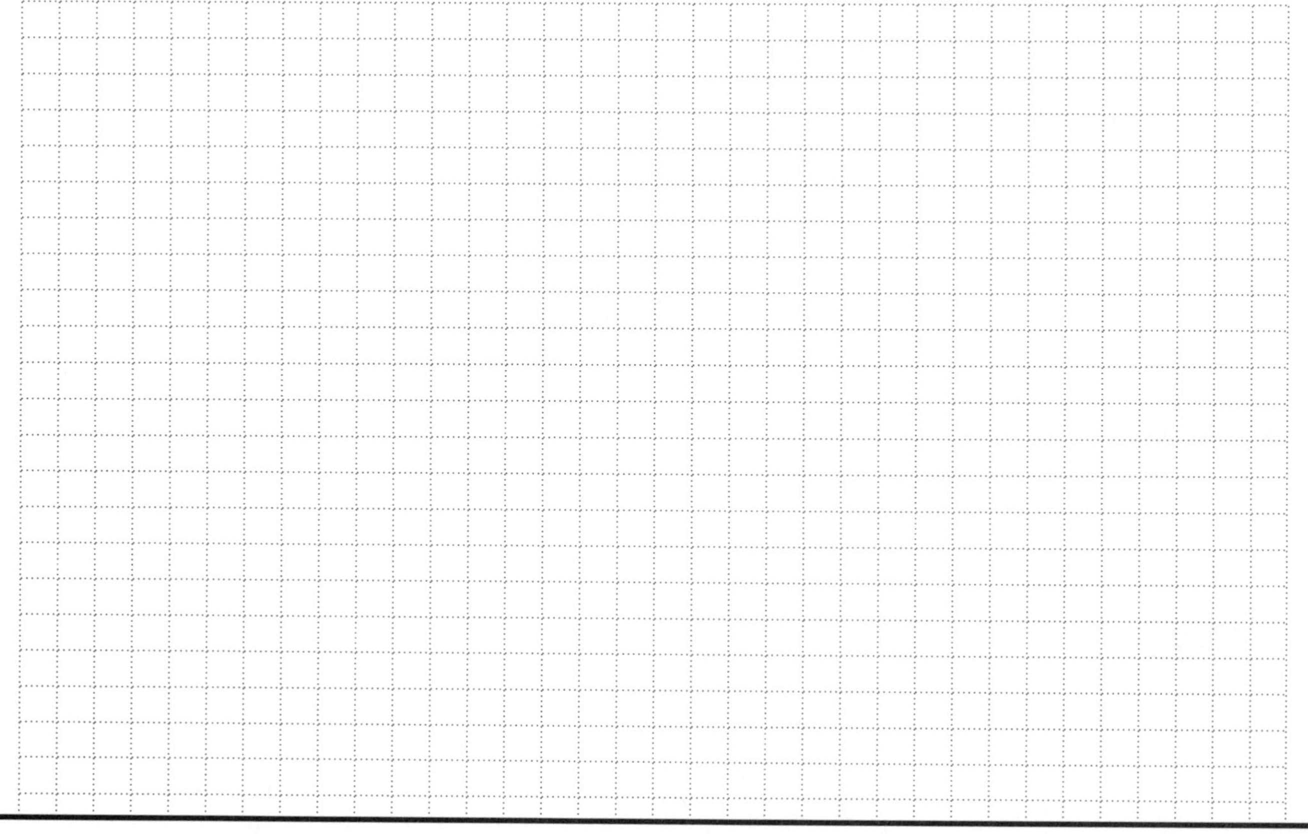

small mass						TABLE B					larger mass
full swings	distance rolled (to nearest cm)			average distance rolled in this time interval	distance at this time interval minus distance of last time interval	full swings	distance rolled (to nearest cm)			average distance rolled in this time interval	distance at this time interval minus distance of last time interval
	1st trial	2nd trial	3rd trial				1st trial	2nd trial	3rd trial		

Compare the numbers in this column →

Compare the numbers in this column →

How much farther did the object roll at each time interval compared to the distance it rolled in the previous time interval? (Your answer is the acceleration of the ball at this angle of incline.)

PS9 Lesson #4

GALILEO, NEWTON, AND THE LAW OF GRAVITY

Work Date: ___/___/___

LESSON OBJECTIVE

Students will find the force of gravity between two massive objects using Sir Isaac Newton's formula for calculating gravitational force.

Classroom Activities

On Your Mark!

Prompt discussion. Given the results of the activities performed in the last two lessons, how did the earth "know" to pull down the heavier object with greater force than the lighter one so that both would hit the ground at the same time? Answer: The force of gravity must depend on the masses of both objects.

Get Set!

Introduce Newton's formula for calculating the force of gravity between two objects: $\mathbf{F_g = G(m_1 \times m_2) \div r^2}$. Explain that increasing the values in the numerator (i.e., $m_1 \times m_2$), increases the force between the objects. Increasing the value in the denominator (i.e., r^2) decreases the force between the objects. Write common fractions on the board to illustrate this point (i.e., $\frac{3}{5} > \frac{2}{5}$; but $\frac{1}{4} < \frac{1}{2}$). Draw the illustration below on the board to explain the meaning of the gravitational constant. Each sphere has a mass of 1 kilogram. They are placed 1 meter apart. The force between the two spheres can be measured to be 6.67×10^{-11} newtons. The force of gravity between any two objects is a multiple of this constant value. Ask: What would life be like if the universal gravitational constant were one-half or one-tenth its actual value?

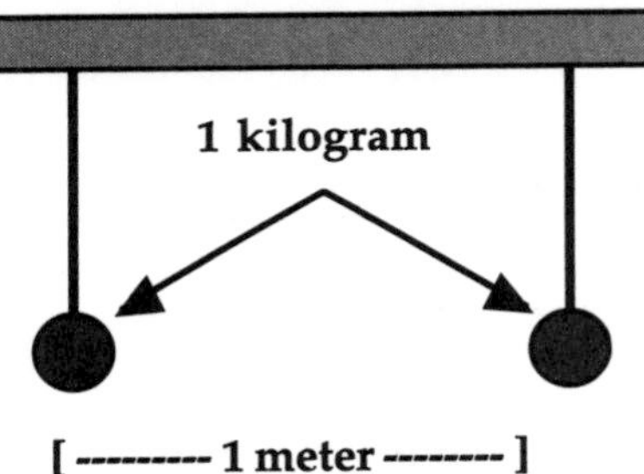

Go!

Have students complete the problems on Journal Sheet #4.

Materials

Journal Sheet #4

ANSWERS TO PROBLEMS OF JOURNAL SHEET #4

1. $\approx 3 \times 10^{-10}$ newtons: a very small force.
2. 300 newtons. It would weigh less on earth where the acceleration due to gravity near the surface is only 9.8 m/s^2.
3. Zero. The astronaut is "weightless." His forward momentum is matched by the pull of earth's gravity.

Name: ______________________________ Period:________ Date: ___/___/___

PS9 Journal Sheet #4

GALILEO, NEWTON, AND THE LAW OF GRAVITY

PROBLEMS

Directions: **Work out each problem in the open space provided on this Journal Sheet. Show all formulas and circle the final answer to each problem.**

1. What is the gravitational force between an asteroid with a mass of 1.2×10^7 kilograms and a spacecraft with a mass of 3.6×10^3 kilograms when the space craft is 100,000 meters from the asteroid?

2. A 25-kilogram space probe near the surface of a planet accelerates toward that surface at 12 m/s^2. What would the object weigh near the surface of this planet? Would the space probe weigh more or less on earth?

3. A 60-kilogram astronaut is asleep aboard the shuttle in orbit 200 kilometers above the Pacific Ocean. The shuttle is moving at approximately 27,000 kilometers per hour, the velocity needed to keep it in orbit. How much does the astronaut weigh?

Name: ______________________ **Period:**________ **Date:** ___/___/___

PS9 Review Quiz

Directions: Keep your eyes on your own work.
Read all directions and questions carefully.
THINK BEFORE YOU ANSWER!
Watch your spelling, be neat, and do the best you can.

CLASSWORK	(~40):	______
HOMEWORK	(~20):	______
CURRENT EVENT	(~10):	______
TEST	(~30):	______
TOTAL	(~100):	______

($A \geq 90$, $B \geq 80$, $C \geq 70$, $D \geq 60$, $F < 60$)

LETTER GRADE: ______

TEACHER'S COMMENTS: ______________________________

GALILEO, NEWTON, AND THE LAW OF GRAVITY

TRUE–FALSE FILL-IN: If the statement is true, write the word TRUE. If the statement is false, change the underlined word to make the statement true. *10 points*

__________ 1. The Italian scientist <u>Galileo</u> dropped objects of differing mass from the Leaning Tower of Pisa.

__________ 2. If there were no atmosphere, objects dropped from the same height at the same instant would hit the ground at <u>the same</u> time(s).

__________ 3. <u>Sir Isaac Newton</u> based his Universal Law of Gravitation on the work of Galileo.

__________ 4. Newton explained that <u>all</u> matter has gravity.

__________ 5. The force of gravity between two objects <u>decreases</u> as the distance between them increases.

__________ 6. The force of the earth's gravitational pull on an object near its surface is called <u>mass</u>.

__________ 7. Weight and mass <u>are</u> the same thing.

__________ 8. A 50-kilogram object near the surface of a planet that accelerates objects toward its surface at 10 meters/second/second would weigh <u>5 newtons</u>.

__________ 9. A 100-gram object near the surface of a planet that accelerates objects toward its surface at 10 centimeters/second/second would weigh <u>1000 dynes</u>.

__________ 10. A falling object <u>does not</u> have weight.

PS9 Review Quiz *(cont'd)*

PROBLEM

Directions: Answer questions #11 through #15 using Figure I.

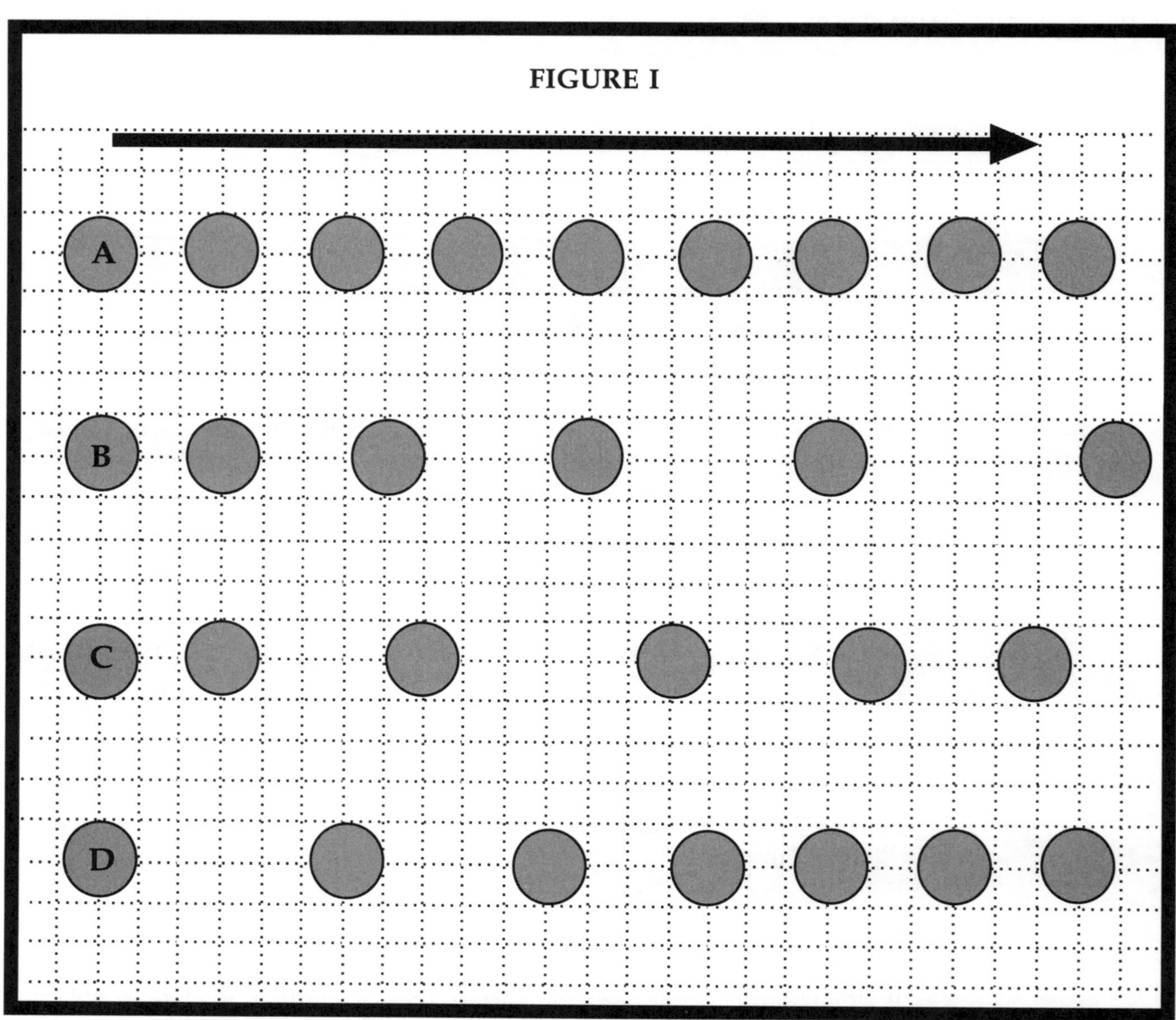

11. Which object is accelerating at a constant rate? ________
12. Which object is moving at a constant velocity? ________
13. Which object is accelerating then decelerating? ________
14. Which object decelerates then moves at a constant velocity? ________
15. Which object could be in a gravitational field? ________

________________________ Student's Signature

________________________ Parent's Signature

____/____/____ Date

PS10

ARCHIMEDES AND BERNOULLI: FLUID PRESSURE

TEACHER'S CLASSWORK AGENDA AND CONTENT NOTES

Classwork Agenda for the Week

1. Students will demonstrate how pressure is related to force.
2. Students will demonstrate that atmospheric pressure can change the shape of objects.
3. Students will demonstrate how to use Bernoulli's Principle to explain how flight is possible.
4. Students will demonstrate how to use Archimedes' Principle to explain why some objects float while others sink.

Content Notes for Lecture and Discussion

The concept of pressure is frequently confused with that of force. A **force** is a push or pull on an object which, according to **Newton's Second Law of Motion** can be summarized in the formula, $\mathbf{f = m \times a}$. **Pressure**, on the other hand, is the force exerted by a solid or fluid over a given surface area. Pressure is a measure of force over area: $\mathbf{P = F \div A}$. Lesson #1 includes several demonstrations that illustrate the subtle difference between these two related concepts.

In Lesson #2, students will become familiar with the idea that the atmosphere, like any other fluid, exerts pressure on adjacent surfaces. Since the atmosphere rises to about 500 kilometers above the earth's surface, a column of air measuring a mere inch in diameter weighs approximately 14.7 pounds at sea level. In metric units of measure, this is equal to about one million (10^6) dynes per square centimeter. This is equal to the weight of approximately 1,000 paper clips exerted over an area of one square centimeter. This measure of atmospheric pressure is called one **bar**. The Italian scientist, **Evangelista Torricelli** (b. 1608; d. 1647) designed the first **barometer** to measure the force being exerted by the atmosphere. At the same time, he gained notoriety by demonstrating the existence of a **vacuum** which **Aristotle** (b. 384 B.C.; d. 322 B.C.) had claimed was an impossible notion. Torricelli was intrigued by the fact that the liquid contents of an inverted test tube did not drain when placed in an open cup of liquid. He reasoned that the force of the atmosphere on the surface of the liquid in the open cup held up the liquid in the test tube. Upon taking his barometer to higher elevations Torricelli discovered that the liquid in the test tube did drop. Since no air was present in the test tube, the space created at the top of the test tube must be completely empty. In addition to measuring the extent to which atmospheric pressure dropped at higher elevations, Torricelli had created a material vacuum. Lesson #2 includes several demonstrations and activities that illustrate the effects of atmospheric pressure and the principle of a simple **liquid barometer**.

Daniel Bernoulli (b. 1700; d. 1782) is the most renowned member of a once distinguished family of Swiss mathematicians. In 1738, he published his most famous book, *Hydrodynamica*, in which he examined the properties of flowing gases, particularly their pressure, density, and speed of flow. In that volume, Bernoulli established the relationship between pressure, density, and the speed at which a fluid flows in his famous law: **Bernoulli's Principle**. Bernoulli discovered that the more rapidly a fluid flows across a surface the less pressure it exerts on that surface. **The Wright Brothers** used this idea to design the wings of the first propeller-driven airplane and successfully carried a man aloft in motorized flight. Lesson #3 includes several activities which demonstrate this basic physical principle. Daniel Bernoulli's work also laid the foundation of the kinetic theory of gases by establishing the statistical relationship between the randomly moving molecules of a gas at a particular temperature and the pressure the gas exerts on surrounding surfaces.

There is a popular myth that the brilliant mathematician and engineer, **Archimedes** (b. 287 B.C.; d. 212 B.C.), leaped from his bath and ran screaming naked into the street: "Eureka! Eureka!" He cried. "I have found it! I have found it!" What he had realized was that a body immersed in a fluid displaces that fluid and that the weight of the fluid displaced must equal the weight of the body itself. He concluded that in order for the body to float, it must displace a weight of fluid equal to its own weight. He summarized his idea in the following way: A body immersed in a fluid will lose a weight equal to the weight of the fluid it displaces. This simple statement is known as **Archimedes' Principle**. Archimedes later refined his idea, realizing that the **density** of an object determines whether or not the object will float or sink. Density is mass per unit volume. If an object is less dense than the fluid into which it is placed, then it will float. If it is not, then the object will sink. Since a **fluid** is anything that flows, either liquid or gas, this principle holds for both water and air. Archimedes' Principle explains not only why heavy ships float on water while a paper clip sinks but also why hot air balloons float on air while a feather falls to earth. The upward pressure exerted by molecules of water or air is called **buoyancy**. Lesson #4 includes several activities that demonstrate why some objects float while others sink.

ANSWERS TO THE HOMEWORK PROBLEMS

1. P = F ÷ A
   = 150 newtons ÷ (0.5 m)(0.5 m) = 150 newtons ÷ 0.25 m^2
   = 600 newtons/m^2
2. D = m ÷ v
   = (400 grams of marbles + 75-gram boat) ÷ 500 milliliters water
   = 475 ÷ 500
   = 0.95 grams/ml which is less than the density of water (1 gram/ml)
   The boat filled with marbles will float.
3. Drawings should show the column of liquid in the barometer falling with a decrease in atmospheric pressure and rising with an increase in atmospheric pressure.
4. Answers will vary but should include that fact that the weight, and therefore the force per unit area, of the water above a diver increases with depth, exerting more pressure on the diver's sensitive eardrums.

ANSWERS TO THE END-OF-THE-WEEK REVIEW QUIZ

1. fluid
2. true
3. true
4. true
5. true
6. barometer
7. less
8. lose
9. volume
10. mass
11. nothing (it's density that's important)
12. buoyancy
13. dividing
14. true
15. sometimes

PROBLEM

Solutions may vary but should have the basic elements shown in this solution.

Density of the planet's liquid
D = m ÷ v
= 200 grams ÷ 50 ml
= 4 grams/ml

Density of the spacecraft
D = m ÷ v
= 100,000,000 grams ÷ 20,000,000 ml
= 5 grams/ml

16. The density of the spacecraft is greater than the density of the planet's sticky surface. If the engine is turned off, the spacecraft will sink!

Name: ______________________ Period: ________ Date: ___/___/___

PS10 FACT SHEET

ARCHIMEDES AND BERNOULLI: FLUID PRESSURE

CLASSWORK AGENDA FOR THE WEEK

(1) Show how pressure is related to force.
(3) Show how atmospheric pressure can change the shape of objects.
(4) Use Bernoulli's Principle to explain how flight is possible.
(5) Use Archimedes' Principle to explain why some objects float while others sink.

According to the **Atomic–Molecular Theory of Matter**, all matter is made up of tiny particles that are in constant motion. All **solids**, **liquids**, and **gases** are matter and, therefore, are made of tiny particles in motion. A **fluid** is defined as anything that flows. Both liquids and gases are fluids. The particles in a fluid exert **pressure** on the surfaces they touch. The tiny particles that comprise matter make trillions of collisions per second against nearby surfaces. The total pressure exerted on a surface by those collisions can be expressed as follows:

$$\mathbf{P = F \div A}$$

where **P** is the pressure on the surface, **F** is the force exerted by a solid or fluid against another surface, and **A** is the **area** over which the force is exerted.

The atmosphere is composed several kinds of gas. At sea level, the **atmospheric pressure** exerted by those gases is about 14.7 pounds per square inch. Atmospheric pressure at sea level is called one bar (or 10^6 dynes per square centimeter). A **millibar** is one thousandth of a bar. A device used to measure air pressure is called a **barometer**. It is a simple tool consisting of an inverted tube filled with a liquid such as water or mercury. The atmospheric pressure on the surface of the liquid surrounding the inverted tube keeps the liquid in the tube from falling. When atmospheric pressure decreases, the level of the liquid in the tube falls. An increase in atmospheric pressure causes the level of the liquid in the tube to rise.

The Swiss scientist, **Daniel Bernoulli** (b. 1700; d. 1782), discovered that the more rapidly a fluid moves across a surface the less force it exerts on that surface. **The Wright Brothers** used this idea, called **Bernoulli's Principle**, to design the wings of the first propeller-driven airplane to successfully carry a man in motorized flight.

The Greek scientist, **Archimedes** (b. 287 B.C.; d. 212 B.C.), discovered that an object will float in a fluid if it can displace (or "push away") a greater weight in fluid than the object itself weighs. This idea is known as **Archimedes' Principle**. It can be stated as follows: An object immersed in a fluid will lose a weight equal to the volume of fluid it displaces. The upward push of the water against an object with less density than the water is called **buoyancy**.

Homework Directions

SHOW ALL MATHEMATICAL FORMULAS AND CALCULATIONS IN SOLVING PROBLEMS #1 and #2. BE SURE TO INCLUDE CORRECT UNITS OF MEASURE WITH YOUR ANSWER.

1. What is the pressure exerted on a table by a metal block weighing 150 newtons if the bottom of the block measures 0.5 meters square?

2. John built a plastic model boat that could displace 500 milliters of water. He loaded the boat with 100 marbles, each with a mass of 4 grams. The boat itself had a mass of 75 grams. Will John's boat float or sink when loaded with the marbles?

Assignment due: _________

3. Draw and label the parts of a liquid barometer. Draw what would happen if the atmospheric pressure surrounding the barometer suddenly decreased. Draw what would happen if the atmospheric pressure surrounding the barometer suddenly increased.

4. In 20 words or less, explain why your ears hurt when you dive to the bottom of a deep swimming pool.

Assignment due: _________

______________________ Student's Signature

______________________ Parent's Signature

____/____/____ Date

PS10 Lesson #1

ARCHIMEDES AND BERNOULLI: FLUID PRESSURE

Work Date: ___/___/___

LESSON OBJECTIVE

Students will demonstrate how pressure is related to force.

Classroom Activities

On Your Mark!

Open discussion by asking, "What is the difference between force and pressure?" Review the concept of force as summarized in **Newton's Second Law of Motion**: **f = m × a**. Perform the first demonstration described below to illustrate that *pressure* is force per unit area: **P = F ÷ A**. Work out one or two problems on the board to show students how to calculate the pressure exerted by a solid resting on a surface.

Get Set!

Perform the demonstration illustrated below: (1) Place a heavy textbook on a thin piece of wood and place the wood on top of a piece of packaging styrofoam like that used to pack a television or computer monitor. Have students agree that the wood and textbook are exerting a force on the styrofoam. (2) Drive a nail through the piece of wood. (3) Place the textbook on top of the wood as before and hold them over the styrofoam with the point of the nail pointing down. (4) Ask: "What will happen when the nail is placed down on the styrofoam?" Students will probably predict that the nail will drive a hole into the styrofoam. (5) Do it, then ask: "Was it the small added weight of the nail that caused this to happen? Or, was the force behind the nail more concentrated this time?" Conclude that the pressure at the end of the nail was much greater than the pressure exerted on the styrofoam by the wood alone because the force was concentrated in a much smaller area (i.e., at the point of the nail).

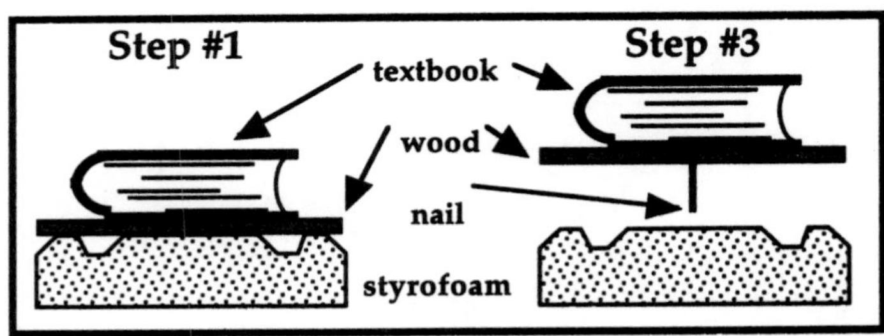

Go!

Have students measure the pressure exerted on their table tops by wooden blocks or sand-filled cereal boxes as described in the activity on Journal Sheet #1. Have them record their results in Table A.

Materials

wooden blocks/sand-filled cereal boxes, balances, metric rulers, slab of wood, nail, hammer, heavy textbook, molded styrofoam

Name: ________________________ Period:________ Date: ___/___/___

PS10 Journal Sheet #1

ARCHIMEDES AND BERNOULLI: FLUID PRESSURE

TABLE A

Directions: Number the sides of your block of wood or sand-filled cereal box and place it on end on your table top. Calculate the pressure exerted on your table top by that side of the block or box. Would the pressure exerted on the table top be the same if another side of the box were placed on the table? Find out by completing the table below. NOTE: The unit of measure for pressure will be in dynes per square centimeter.

object side	force*	area**	pressure

* force = weight or $w = m \times g$; where $g = 980\ cm/s^2$

** area = $l \times w$

PS10 Lesson #2

ARCHIMEDES AND BERNOULLI: FLUID PRESSURE

Work Date: ____/____/____

LESSON OBJECTIVE

Students will demonstrate that atmospheric pressure can change the shape of objects.

Classroom Activities

On Your Mark!

Begin discussion by reminding students that air is made up of several gases: 78% nitrogen, 21% oxygen, 1% carbon dioxide and other substances. Explain that these gases are made of molecules which are made of atoms that have mass and, according to the Atomic-Molecular Theory of Matter, are constantly moving around.

Get Set!

Perform the demonstration shown in Illustration A: (1) Fill a test tube with water and place a playing/index card over the top. (2) Slowly invert the test tube over a sink or pan while holding the card in place. With practice, you should be able to keep the card from falling. The pressure of the atmosphere will hold it in place as illustrated below. Ask: "What force is holding up the card against the force of gravity trying to pull the water down?" Perform the demonstration shown in Illustration B: (1) Invert a water-filled test tube into a beaker/glass of water allowing a small amount of water to spill from the tube, creating a space filled with air inside the tube. (2) Ask: "What will happen to the water in the tube if I start to lift the test tube off the bottom of the beaker?" Do it. Students will observe that the level of the water in the tube does not change. Draw Illustration B on the board to show the forces at work in this "liquid barometer." Describe the experiments of **Evangelista Torricelli** mentioned in the Teachers Agenda and Notes for Class Discussion. Perform the demonstration shown in Illustration C: (1) Fill a 2-liter plastic bottle with water. (2) Cork it with a single-holed stopper connected to a 6-foot length of rubber tubing. (3) Cork a second 2-liter plastic bottle with a double-holed rubber stopper connected to the opposite end of the tubing. (4) Place the empty bottle on the floor, invert the filled bottle, and raise it over the bottle on the floor high over your head. As the water drains from the top bottle into the one on the floor, the atmosphere will crush the bottle in your hand.

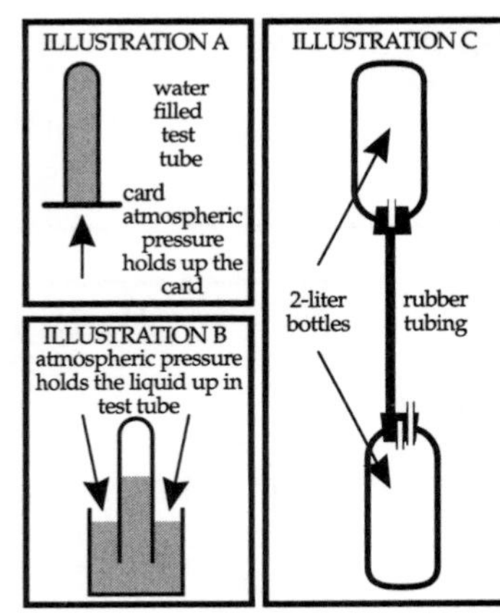

Go!

Have students perform the activity described in Figure B on Journal Sheet #2. Explain that heating the can caused the air inside to expand and escape. When the can was inverted into the bowl of cool water, the remaining air inside contracted, allowing atmospheric pressure outside the can to crush it.

Materials

test tubes, beakers, index/playing cards, two 2-liter plastic bottles, rubber tubing, one single-holed stopper, one double-holed stopper, soda cans, Bunsen burners, ring stands and clamp, tong

Name: ______________________ Period:________ Date: ___/___/___

PS10 JOURNAL SHEET #2

ARCHIMEDES AND BERNOULLI: FLUID PRESSURE

FIGURE B

Directions: (1) Pour several milliliters of water into a soda can. (2) Place the soda can onto a ring stand and secure it loosely with another ring and clamp so that it cannot be toppled but can be removed easily with tongs. (3) Fill a small bowl with water. (4) Turn on the Bunsen burner and wait until steam begins to escape from the can. (5) Count slowly to 20. (6) Holding the can securely with tongs, quickly lift and invert the can upside down into the bowl of cool water.

GENERAL SAFETY PRECAUTIONS

Be sure you are familiar with the proper use of a Bunsen burner. Wear goggles. Do not touch any part of the equipment without heat-resistant gloves or tongs.

Perform the step-by-step demonstration in FIGURE B and fill in the Reports #1, #2, and #3.

#1. Report what you observed when you put the hot can upside down into the water. ______________________

#2 Explain why this happened. ______________________

#3. Why was it necessary to heat the can before you placed it into the bowl? ______________________

PS10 Lesson #3

ARCHIMEDES AND BERNOULLI: FLUID PRESSURE

Work Date: ____/____/____

LESSON OBJECTIVE

Students will demonstrate how to use Bernoulli's Principle to explain how flight is possible.

Classroom Activities

On Your Mark!

Tell students to take out a plain piece of looseleaf paper and construct a simple paper airplane. Have them spend a few fun moments challenging one another to see whose model will fly the farthest. *Take precautions to make sure no one gets poked in the eye with one!* At the end of the brief contest have students examine the wings of their craft and write down what facet of the wing's structure they think helped their model to stay aloft. Ask: "Was it the size of the wing? Was it the thickness or angle the wing made with the model's fuselage?"

Get Set!

Place a ping-pong ball in a funnel. Attach a piece of rubber tubing to the narrow end of the funnel. Hold it up and turn it over to show students how easily the ball falls out. Then place the ball back into the funnel, hold it up, and challenge someone to blow the ball out of the funnel by exhaling into the rubber tube as shown in Illustration D. It cannot be done! Next: Place the ball on a table and hold the funnel over it as shown in Illustration E. Blow forcefully into the tube. With practice you will be able to "suck" the ball into the funnel as though it were a vacuum cleaner. Explain **Bernoulli's Principle** and have students copy it on Journal Sheet #3: The more rapidly a fluid (i.e., liquid or gas) moves across a surface, the less pressure it exerts on that surface. Ask students to discuss how Bernoulli's Principle applies to the previous demonstrations. For extra credit, ask students to draw the inside of a vacuum cleaner to show how it works.

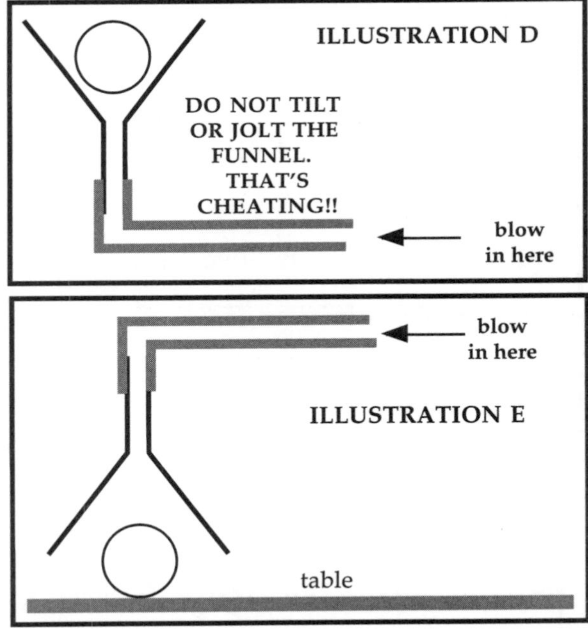

Go!

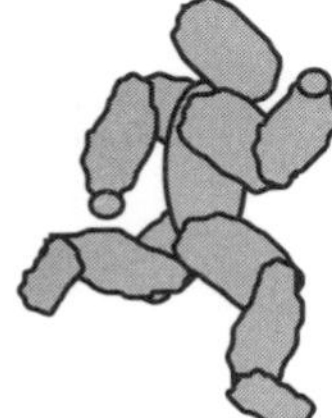

Have students perform the activity described in Figure C on Journal Sheet 3#.

Materials

funnels, rubber tubing, ping pong balls, index card

Name: ______________________ Period:________ Date: ___/___/___

PS10 JOURNAL SHEET #3

ARCHIMEDES AND BERNOULLI: FLUID PRESSURE

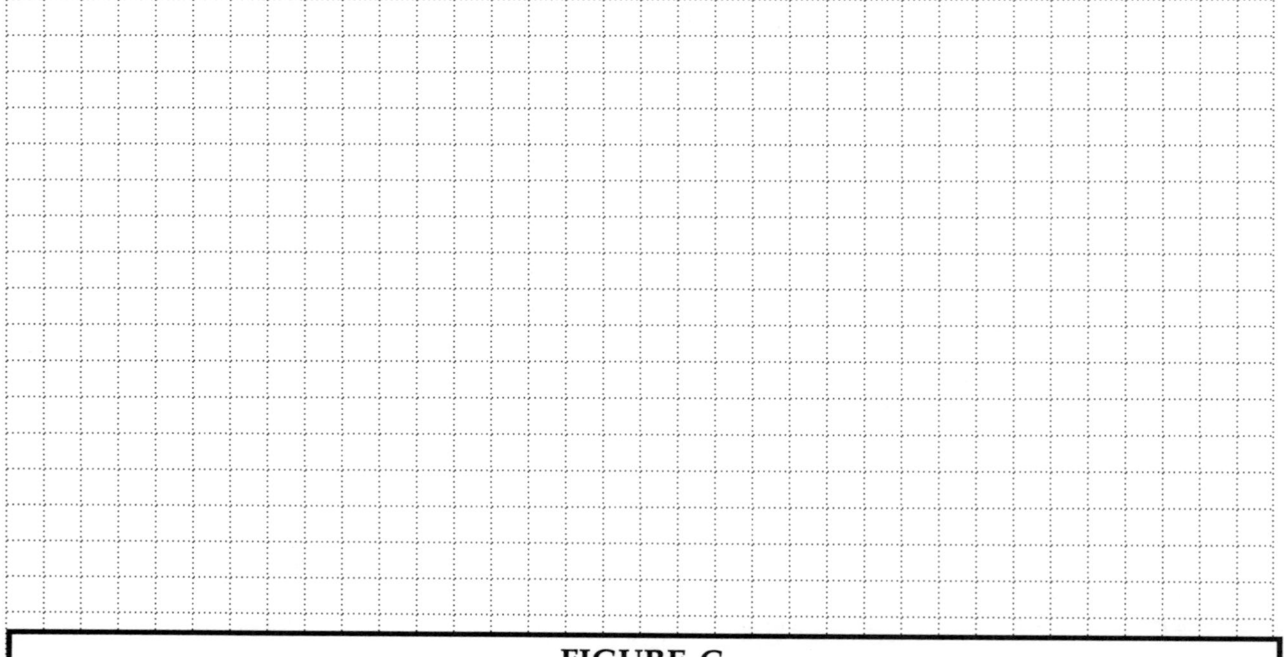

FIGURE C

(1) Fold an index card as shown at right. (2) Place it on a table. (3) Try to flip it over by blowing as hard as you can underneath it.

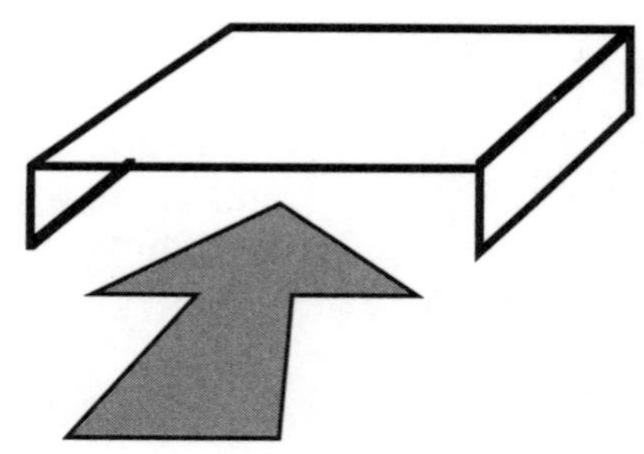

Describe what you observed. ______________________

____

Why did the card behave the way it did?______________________

____

CROSS SECTION OF AN AIRPLANE WING

The air flowing over the wing must travel a farther distance in the same amount of time than the air flowing under the wing. In doing so, the "overflow" exerts less pressure on the top of the wing than the "underflow" and atmospheric pressure exert on the bottom of the wing. This creates "lift" which pushes the wing and aircraft up into the air.

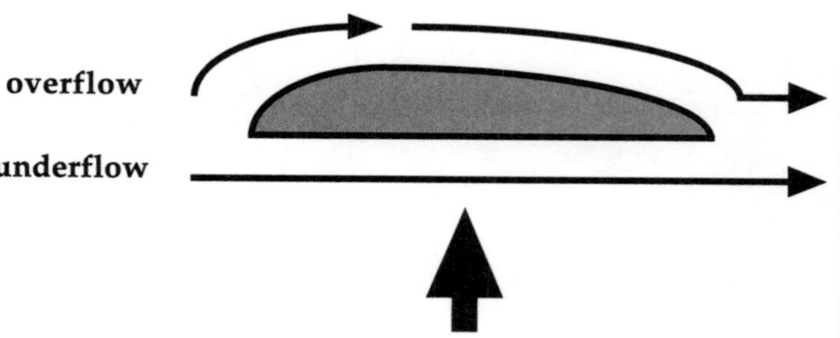

PS10 Lesson #4

ARCHIMEDES AND BERNOULLI: FLUID PRESSURE

Work Date: ____/____/____

LESSON OBJECTIVE

Students will demonstrate how to use Archimedes' Principle to explain why some objects float while others sink.

Classroom Activities

On Your Mark!

Begin discussion by asking the following question: "Why was a ship as large as the *Titanic* able to float while a penny sinks?" If students suggest that "weight" has something to do with whether or not an object floats, point out that the *Titanic* had a mass of 60 million kilograms while a penny weighs a mere 3 grams (0.003 kg). Students will eventually suggest that the shape of the object has something to do with whether or not it will float. Tell the legendary story of Archimedes (b. 287 B.C.; d. 212 B.C.) mentioned in the Teacher's Classwork Agenda and Content Notes and have students copy **Archimedes' Principle** into their notes on Journal Sheet #4.

Get Set!

Assist students in completing the activity described in Figure D. Then, challenge them to float 20 pennies in a pan of water by constructing a small rectangular "boat" out of construction paper. They should plan their boat by figuring out the minimum volume needed to float the pennies. Have them consider these facts: (1) a single penny has a mass of about 3 grams, (2) water has a mass of 1 gram per cubic centimeter (or milliliter), (3) the paper boat itself has a mass of a few grams. Refer them to the box of Construction Tips on Journal Sheet #4.

Go!

Have students perform the activity described in Figure D and fill in Table B on Journal Sheet #4. Set up a water pan on a convenient counter so that students can test their paper boats. The winner is the person or group with the smallest boat able to float the 20 pennies. Considering the facts given above, a boat with a volume of approximately 70 cubic centimeters should do the job. (Volume = length × width × height)

Materials

graduated cylinders, brass weights, balance, pan, index cards, tape/staples

Name: ______________________________ Period:________ Date: ____/____/____

PS10 Journal Sheet #4

ARCHIMEDES AND BERNOULLI: FLUID PRESSURE

TABLE B		
mass of 100-ml cylinder	mass of weights	ml water displaced

FIGURE D

Directions: (1) Fill a 500-ml graduated cylinder with 250 ml of water. (2) Find the mass of a smaller 100-ml graduated cylinder using a balance. (3) Place the small cylinder into the larger one and record the amount of water displaced by the small cylinder. (4) Remove the small cylinder and place a 20-gram brass weight into the cylinder. (5) Repeat step #3. (6) Remove the small cylinder and repeat the procedure to find out the amount of water displaced when the cylinder holds 30 grams, 40 grams, and 50 grams of mass.

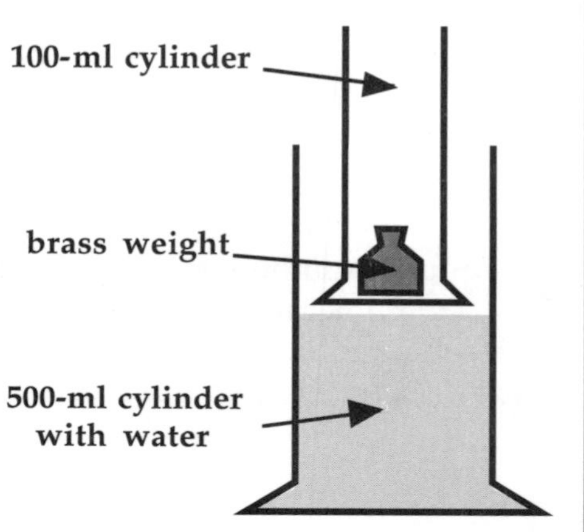

CONSTRUCTION TIPS

After figuring the correct dimensions of your boat, start with a piece of paper that looks like Drawing #1. Fold in on the dotted lines as shown in Drawing #2 and tape the corners of your boat closed. This will prevent leaks from cuts or staples.

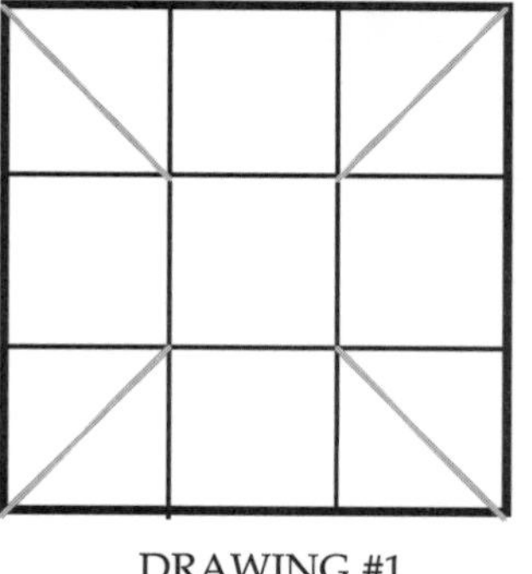

DRAWING #1

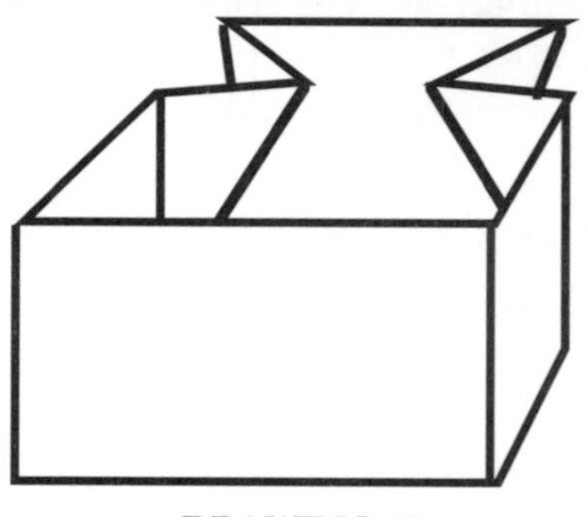

DRAWING #2

Name: ______________________ Period:________ Date: ____/____/____

PS10 Review Quiz

Directions: Keep your eyes on your own work.
Read all directions and questions carefully.
THINK BEFORE YOU ANSWER!
Watch your spelling, be neat, and do the best you can.

CLASSWORK	(~40):	______
HOMEWORK	(~20):	______
CURRENT EVENT	(~10):	______
TEST	(~30):	______
TOTAL	(~100):	______

(A ≥ 90, B ≥ 80, C ≥ 70, D ≥ 60, F < 60)

LETTER GRADE: ______

TEACHER'S COMMENTS: ____

ARCHIMEDES AND BERNOULLI: FLUID PRESSURE

TRUE–FALSE FILL-IN: If the statement is true, write the word TRUE. If the statement is false, change the underlined word to make the statement true. *15 points*

______________ 1. Any substance that flows is called a(n) <u>liquid</u>.

______________ 2. <u>All</u> fluids exert pressure on the objects they touch.

______________ 3. Pressure is force <u>divided</u> by area.

______________ 4. <u>All</u> atoms and molecules have small amounts of mass.

______________ 5. Atmospheric pressure at sea level is about <u>14.7</u> pounds per square inch.

______________ 6. A device used to measure air pressure is a(n) <u>thermometer</u>.

______________ 7. Bernoulli discovered that the faster a fluid moves across a surface the <u>more</u> force it exerts on that surface.

______________ 8. Archimedes discovered that an object placed in a fluid will <u>gain</u> a weight equal to the volume of fluid it displaces.

______________ 9. <u>Mass</u> is the amount of space occupied by an object.

______________ 10. <u>Volume</u> is the amount of matter in an object.

______________ 11. An object's weight has <u>a lot</u> to do with its ability to float.

______________ 12. The upward force of a fluid is called <u>density</u>.

______________ 13. The density of an object is found by <u>multiplying</u> its mass by its volume.

______________ 14. An object whose density is <u>less</u> than the density of the fluid in which it is placed will float.

______________ 15. Objects that have air in them will <u>always</u> float.

PROBLEM

Directions: Read the following problem carefully. Show all formulas and mathematical calculations that help you solve the problem. Answer question #16 after doing your calculations. Your calculations are worth *13 points*.

You are an explorer in space. Your small spaceship sets down on the liquid surface of an alien planet. To be safe, you keep the ship's engines running until you are able to make some quick measurements. You quickly sample the liquid surface of the planet and discover that 50 milliliters of the sticky liquid has a mass of 200 grams. The mass of your spacecraft is 100,000,000 grams and takes up 20,000,000 cubic centimeters of volume.

What should you do? Will you sink or float if you turn off your engines?

16. Will your spacecraft sink or float if you turn off the engines? Why? ____________________

____

____ *2 points*

______________________ Student's Signature

______________________ Parent's Signature

____/____/____ Date

PS11

FORCE, WORK, AND SIMPLE MACHINES

TEACHER'S CLASSWORK AGENDA AND CONTENT NOTES

Classwork Agenda for the Week

1. Students will explain the scientific meaning of the term "work" and identify two basic kinds of machines.
2. Students will demonstrate how a lever produces a mechanical advantage.
3. Students will demonstrate how an inclined plane produces a mechanical advantage.
4. Students will demonstrate how a pulley produces a mechanical advantage.

Content Notes for Lecture and Discussion

People have used tools and machines for thousands of years in an effort to shape and control their environment. Even our ancestors *homo erectus* and *homo neanderthalensis* used axes and hammers made of wood and stone more than two hundred-thousand years ago. The first nails were used in the Middle East around 3500 B.C., and the first lathes to shape wood and stone pedestals before 3000 B.C. Scissors made of bronze are known to have been used in Europe and Asia as early as 1000 B.C. Some people find it hard to believe that the early Egyptians used simple machines to build the pyramids. But in fact, these monumental structures were erected using **levers** and **inclined planes**, the two basic simple machines from which most all **compound machines** are devised.

In Lesson #1 students will gain an understanding of how the simple common tools they use every day make work easier. They will discuss and discover what machines can and cannot do. It is a common misperception that machines "save work" when in fact they do nothing of the kind. Machines do, however, make work easier by (1) increasing the force of a worker, (2) increasing the speed of a worker, (3) changing the direction of a worker's applied force, and (4) transferring or transforming energy. Lesson #1 will also introduce students to the mathematical formula scientists use to calculate the amount of work done in any given situation.

The concept of "work" means something different to a scientist than it does to the layman. To a scientist, work is the action of a force through a distance: $\mathbf{W} = \mathbf{F} \times \mathbf{D}$ where **W** is work measured in **joules** (MKS) or **ergs** (CGS); **F** is force measured in **newtons** (MKS) or **dynes** (CGS); and **D** is distance measured in **meters** (MKS) or **centimeters** (CGS). In physics, work and **energy** are equivalent physical concepts. Energy is the ability to do work.

In Lesson #2 students are introduced to the **lever** and the concept of **mechanical advantage**. The mechanical advantage of a particular tool or machine is equal to the number of times that tool or machine multiplies a worker's force or speed. Mathematically, the mechanical advantage of a machine is expressed as follows: $\mathbf{M.A.} = \mathbf{F_r} \div \mathbf{F_e}$ where **M.A.** is mechanical advantage, $\mathbf{F_r}$ is **resistance force** (i.e., the load to be moved), and $\mathbf{F_e}$ is the **effort force** (i.e., the force applied by the worker). **Archimedes** (b. 287 B.C.; d. 212 B.C.) once made the following claim: "Give me a lever long enough and a support on which it can rest, and I could move the world." Students will discover that a lever increases a worker's applied force at the expense of distance (i.e., how far the worker must be from the load they are attempting to move).

In Lesson #3 students will discover the advantages of using an **inclined plane** and how to calculate its mechanical advantage.

PS11 Content Notes *(cont'd)*

In Lesson #4, students are challenged to construct a **pulley** system with a mechanical advantage of 5. Some of the first pulley systems were described by **Hero of Alexandria** who lived in the first century A.D. In his book *Mechanics,* Hero described a **compound pulley** like the one shown in Illustration A.

This series of lessons will prepare students to master the concepts of **machine efficiency** and **power** presented in the next unit.

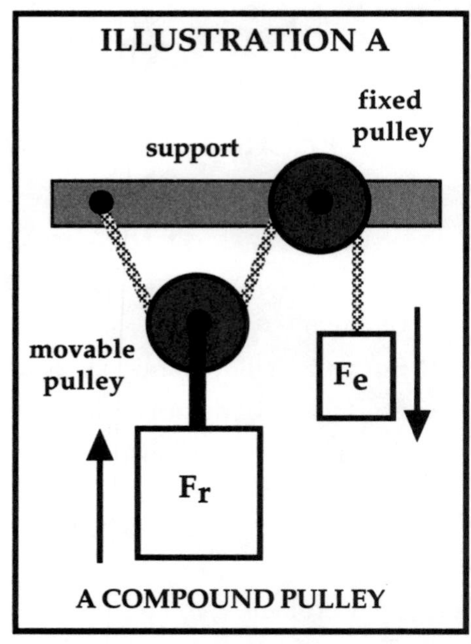

A COMPOUND PULLEY

ANSWERS TO THE HOMEWORK PROBLEMS

1. W = F × D
   = 24 lbs × 16 ft
   = 384 foot-pounds (ft-lbs)
2. D = W ÷ F
   = 120 joules ÷ 60 newtons
   = 2 meters
3. W = F × D = m × a × D; so . . .
   m = W ÷ (a × D)
   = 245 joules ÷ (9.8 m/s^2 × 5 meters)
   = 5 kilograms

ANSWERS TO THE END-OF-THE-WEEK REVIEW QUIZ

1. true	6. never (do not)	(A) increase force
2. work	7. true	(B) increase speed
3. work	8. inclined plane	(C) change the direction of forces
4. true	9. true	(D) transfer or transform energy
5. 18 joules	10. increases	

PROBLEM

W	=	F	×	D		
	=	m	×	a	×	D
16,000 ergs	=	4 grams	×	1,000 cm/s^2	×	?
	=					4 centimeters

Name: ______________________________ Period:________ Date: ____/____/____

PS11 Fact Sheet

FORCE, WORK, AND SIMPLE MACHINES

CLASSWORK AGENDA FOR THE WEEK

(1) Explain the scientific meaning of the term "work" and identify two basic simple machines.
(2) Show how a lever produces a mechanical advantage.
(3) Show how an inclined plane produces a mechanical advantage
(4) Show how a pulley produces a mechanical advantage.

The term **work** means something different to a scientist than it does to most people. To a scientist, work is the action of a force over a distance. According to this definition of work, a person trying but failing to lift a heavy object has not done any work. Perhaps they have burned a few calories; but they have not done any work. In order to do work you must move something. The amount of work done by a worker or machine can be expressed by the following formula:

$$\mathbf{W = F \times D}$$

where **W** is work, **F** is force, and **D** is the distance through which the force is applied.

In the Metric System, a large unit of measure for work (i.e., MKS) is the **newton-meter** or **joule**. The smaller unit of measure in the Metric System (i.e., CGS) is the **dyne-centimeter** or **erg**. So, a person or machine that applies a 2-newton force over a distance of 2 meters has done 4 newton-meters, or 4 joules, of work. In the English System of measurement (i.e., the system that uses inches, feet, and yards), the unit of measure for work is the **foot-pound**. This is because in the English System force is measured in pounds and distance is measured in feet.

A **machine** is a tool that helps us to do work. However, machines do not save work. In order for a machine to function, work or energy must be put into the machine to get it to move an object. This is because the opposing **force of friction** between the moving parts of a machine reduces the amount of work done by that machine. Machines do not save work! They only make work easier.

Machines make work easier by producing a **mechanical advantage**. Machines do this in the following four ways: (1) Machines can increase the force of a worker. (2) Machines can increase the speed of a worker. (3) Machines can change the direction of forces applied by a worker. And, (4) machines can transfer or transform energy.

The mechanical advantage produced by a machine can be expressed as follows:

$$\mathbf{M.A. = F_r \div F_e}$$

where **M.A.** is **mechanical advantage**, $\mathbf{F_r}$ is the **resistance force** the worker must overcome, and $\mathbf{F_e}$ is the **effort force** applied by the worker.

The two basic kinds of **simple machines** are the **lever** and the **inclined plane.** Machines that combine these simple machines are called **compound machines**.

A lever is a bar supported by a **fulcrum** like the one shown in the Figure I. By increasing the distance of the worker's effort force from the fulcrum, a greater force of resistance can be overcome. A **pulley** is a special kind of lever. It is a "continuous lever." A pulley uses a string or cable instead of a rigid bar to support the resistance and effort forces. The mechanical advantage of a pulley is equal to the number of strings or cables supporting the resistance force (i.e., the load).

PS11 Fact Sheet *(cont'd)*

An **inclined plane** is a ramp or tilted surface. An inclined plane produces a mechanical advantage by making it easier to work against the force of gravity. Examples of inclined planes are a **stairway**, a **wedge**, or a **screw**. A screw is a "spiraling" inclined plane. The mechanical advantage of an inclined plane can be calculated by dividing the plane's length by its height.

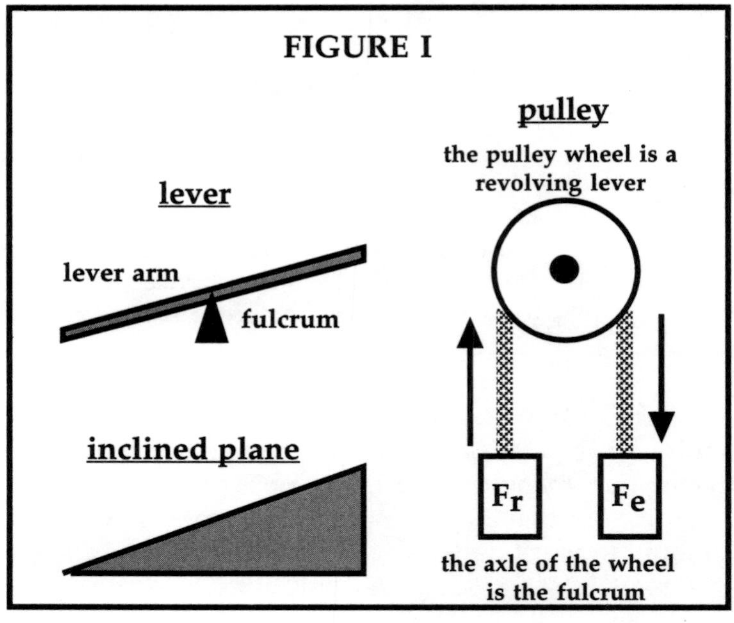

Homework Directions

SHOW ALL MATHEMATICAL FORMULAS AND CALCULATIONS IN SOLVING PROBLEMS #1, #2, AND #3. BE SURE TO INCLUDE CORRECT UNITS OF MEASURE WITH YOUR ANSWER.

1. How much work is done by a person who moves a 24-pound box a distance of 16 feet?
2. How far can a forklift move a 60-newton crate if it can do 120 joules of work?
3. What is the mass of a box that is moved 5 meters by a machine that does 245 joules of work? (*Hint:* You will need to know the rate of acceleration of gravity near the surface of the earth to solve this one.)

Assignment due: ________

________________ Student's Signature

________________ Parent's Signature

___/___/___ Date

PS11 Lesson #1

FORCE, WORK, AND SIMPLE MACHINES

Work Date: ____/____/____

LESSON OBJECTIVE

Students will explain the scientific meaning of the term "work" and identify two basic kinds of machines.

Classroom Activities

On Your Mark!

Prompt students to discuss and list some of the reasons people use tools or simple machines. When they are done, emphasize that tools and machines are used to make work easier. Help students to define the term "work" as a scientist understands it by introducing the following example: A foreman on the job asks one of his workers to move a heavy box. The foreman leaves and the worker tries to move the box but fails to do so. The worker struggles and strains, works up a sweat, but the box does not budge an inch. The foreman returns and frowns. Has the worker done any work? Some students will argue that the worker has done work because he struggled, strained, and worked up a sweat. But in fact no work has been done because the box did not move (i.e., the job was not done). It is true that the worker burned calories, that his heart did work pumping blood through his arteries and veins; but the foreman would hardly be pleased with the result. Explain that the scientific meaning of work requires that a force is applied through a distance: **Work (W) = Force (F) × Distance (D)**. Ask students to look at their list of tools and machines and tell you if any of them "saves work." Point out that the verb "to save" is the opposite of the verb "to spend." All machines must exhaust a certain amount of energy to do their work; so in reality machines never "save energy" which is the ability to do work. Pose the following for discussion: If you earn $5 and you put it in the bank, then you have "saved" $5. But let's say you find a favorite toy on sale for $4 which last week would have cost you $10. If you buy the toy on sale are you really saving any money? Answer: No! You are still spending $4—not saving it! Machines use up energy and "waste work" in the same way people spend money—even when things are on sale. The reason for this is that the **force of friction** always reduces the efficiency of tools and machines. Machines do, however, allow us to spend less energy and do less work in the same way a store sale allows us to spend less money.

Get Set!

Draw the illustrations appearing on the student Fact Sheet and identify all of the parts of a **lever, inclined plane**, and **pulley**. Have students copy your drawing onto Journal Sheet #1. Define the terms **resistance force** (i.e., the load) and **effort force** (i.e., worker's applied force). Explain that the position of the load, effort, and **fulcrum** can be changed to produce the different types of lever as shown in Figure A on Journal Sheet #1. Show students how to calculate the work done in a variety of situations using the formula **W=F×D**. Go over every example you create to make sure that students use the correct units of measure with each calculation.

Go!

After making sure that students feel comfortable with calculating the amount of work done in your mathematical examples, have them complete the activity described in Journal Sheet #1.

Materials

hammer, scissor, bottle opener, etc. for show, Journal Sheet #1

Name: ______________________ Period: ________ Date: ___/___/___

PS11 Journal Sheet #1

FORCE, WORK, AND SIMPLE MACHINES

Draw a pencil, a scissor, a broom, and a hammer. Write the words "load" and "effort" to show where the "load" is placed and where the worker applies their "effort" to do work.

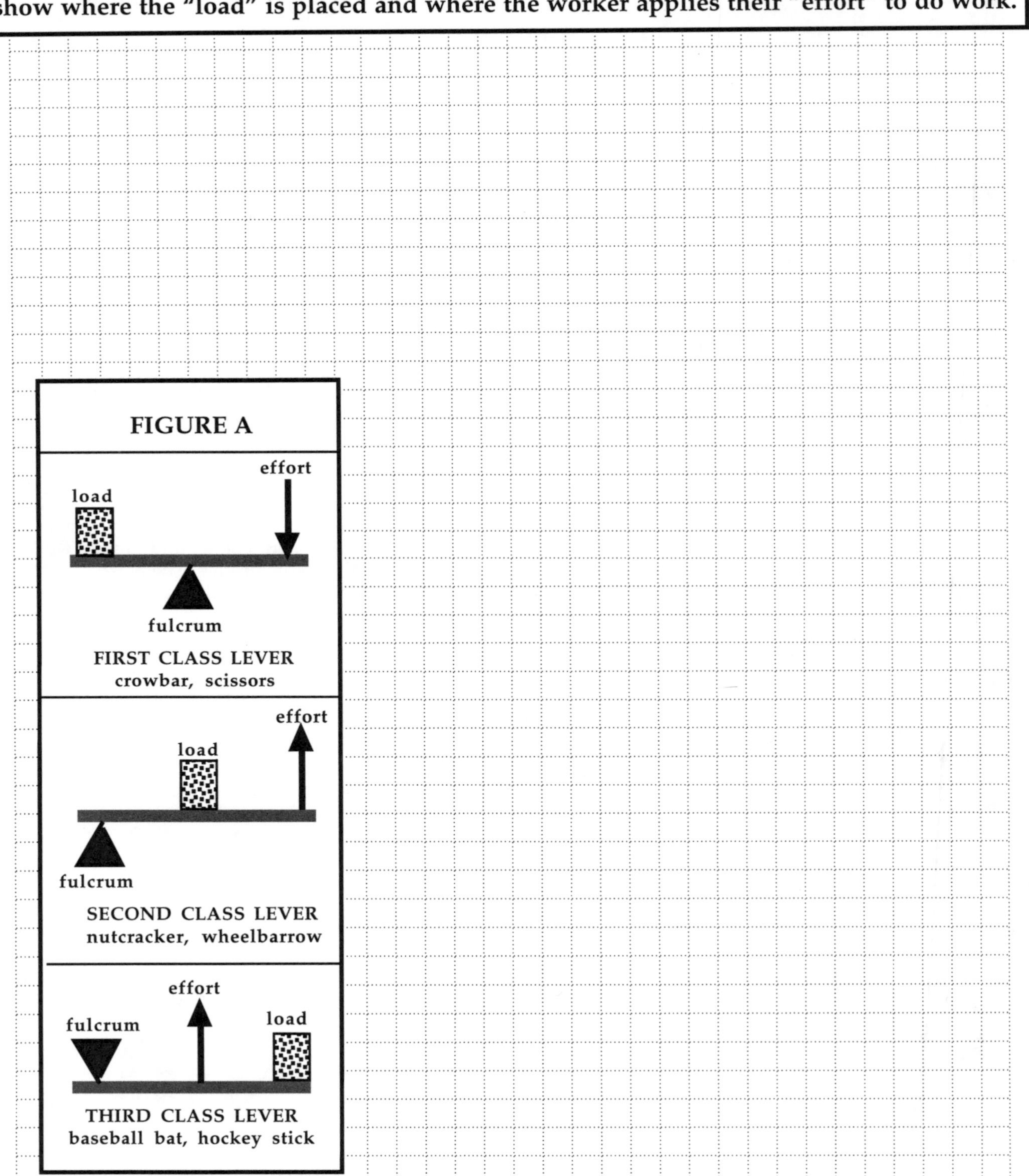

PS11 Lesson #2

FORCE, WORK, AND SIMPLE MACHINES

Work Date: ____/____/____

LESSON OBJECTIVE

Students will demonstrate how a lever produces a mechanical advantage.

Classroom Activities

On Your Mark!

Write Archimedes' famous quote on the board: "Give me a lever long enough and a support on which it can rest, and I could move the world." Ask students to comment on the quote. If we really had an unbreakable lever long enough and a support on which it could rest, could we move the world? Answer: Yes! In reality, of course, no such lever exists. Summarize the basic principle of a lever. The farther from a fulcrum one applies effort to a lever, the less force is required to move a load that is close to the fulcrum.

Get Set!

Put Table A on Journal Sheet #2 on the board and set up your metric ruler and fulcrum as shown. Prepare small wooden triangles before the start of class to act as fulcrums. Balance your ruler on the fulcrum, pointing out that the center of gravity on the ruler may not be exactly at the "50 cm" mark. Tell students that all of their measurements must be taken from the ruler's center of gravity: not the "50 cm" mark. Perform the following demonstration: (1) Place the center of a 50-gram brass weight at the 2 cm mark. Fill in the appropriate box on Table A so that students can see how the table is to be completed. (2) Place a 5-gram brass weight next to the fulcrum opposite the heavier brass weight and ask students to predict when the two weights will balance as you move the lighter weight farther from the fulcrum. (3) Use a pencil to slowly slide the 5-gram mass toward the end of the ruler. When it gets to about 20 cm from the ruler's center of gravity the two masses will balance. (4) Fill out the chart as it appears in Table A, explaining the meaning of each variable: $\mathbf{W_e}$ is the work done by the small mass; $\mathbf{F_e}$ is the force of the small mass (which is equal to $\mathbf{m_e}$, the mass of the small mass, times **a,** the small mass's acceleration due to gravity); and, $\mathbf{D_e}$ is the distance of the small mass from the ruler's center of gravity; $\mathbf{W_r}$ is the work done by the large mass; $\mathbf{F_r}$ is the force of the large mass (which is equal to $\mathbf{m_r}$, the mass of the large mass, times **a,** the large mass's acceleration due to gravity; and, $\mathbf{D_r}$ is the distance of the large mass from the ruler's center of gravity.

Go!

Instruct students to repeat the same activity using different combinations of small and large masses. Make sure they fill out their charts using correct units of measure for each measured quantity.

Materials

metric rulers, brass weights (or pennies at 3 grams/penny), small wooden triangles to use as fulcrums

Name: ______________________ Period: ________ Date: ___/___/___

PS11 Journal Sheet #2

FORCE, WORK, AND SIMPLE MACHINES

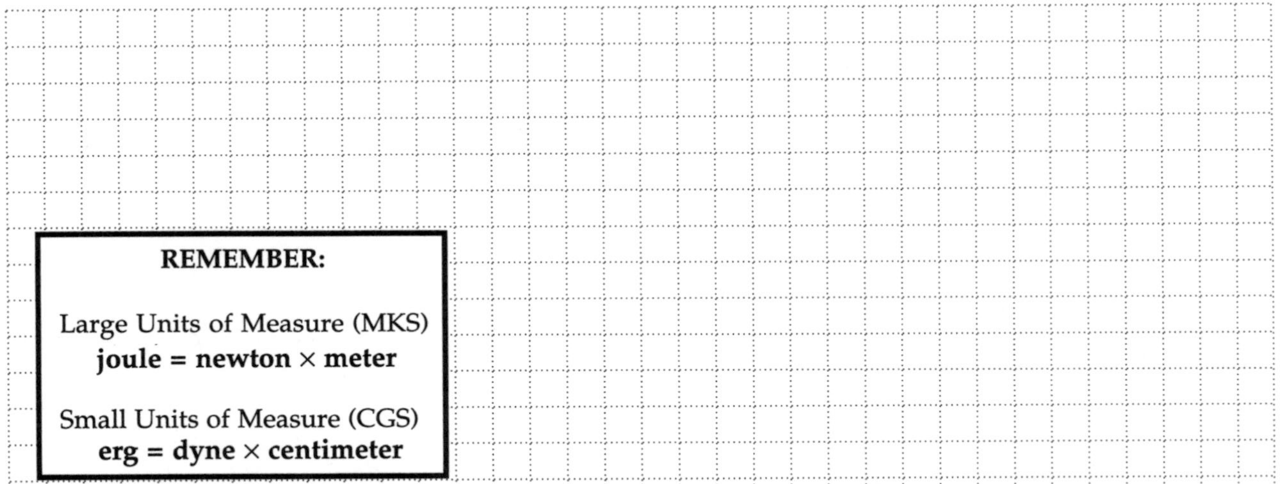

REMEMBER:

Large Units of Measure (MKS)
joule = newton × meter

Small Units of Measure (CGS)
erg = dyne × centimeter

TABLE A

EFFORT = **RESISTANCE**

$W_e = F_e \times D_e = D_r \times F_r = W_r$

(and since **F = m x a** according to Newton's Second Law of Motion)

$W_e = m_e \times a \times D_e = D_r \times m_r \times a = W_r$

W_e	m_e	a	D_e	D_r	m_r	a	W_r
98,000 ergs	5 grams	980 cm/s^2	20 cm	2 cm	50 grams	980 cm/s^2	98,000 ergs

NOTE: In the first example a **5-gram** mass placed **20 centimeters** from the fulcrum will balance a **50-gram** mass placed **2 centimeters** from the fulcrum. With the acceleration of gravity being the same for both masses (**980 cm/s/s**), the work on either side of the fulcrum is **98,000 ergs** (or dyne-centimeters)

FORCE, WORK, AND SIMPLE MACHINES

Work Date: ____/____/____

LESSON OBJECTIVE

Students will demonstrate how an inclined plane produces a mechanical advantage.

Classroom Activities

On Your Mark!

Begin discussion by asking students to list on Journal Sheet #3 tools and machines that have inclined planes (i.e., stairway, knife blade, axe blade, screw or "spiraling inclined plane.") Explain to students that the **mechanical advantage** of an inclined plane can be calculated using the formula $\mathbf{P_l \div P_h = M.A.}$, where $\mathbf{P_l}$ is the length of the plane, $\mathbf{P_h}$ is the height of the plane, and **M.A.** is the mechanical advantage of the plane. Review the formula $\mathbf{F_r \div F_e = M.A.}$ as the general formula for calculating the mechanical advantage of any machine; where $\mathbf{F_r}$ is the resistance force and $\mathbf{F_e}$ is the effort force.

Get Set!

Demonstrate how to construct the set-up shown in Figure B on Journal Sheet #3. Show students how to measure the height of the plane above the surface of the desk. The length of the plane will be 100 cm in every case if they are using standard metric rulers. Before they get going, tell them to obtain accurate measures of mass for their toy car or truck, using a balance.

Go!

Challenge them to balance their toy car or truck on their ramp using a variety of incline heights and a small amount of mass. Make sure they record their measurements and calculations on Table B for every balance test.

Materials

metric rulers, brass weights (or pennies at 3 grams/penny), string, single pulleys, paper cups, ring stands, clamps, table, toy cars and trucks, balance

Name: ______________________ Period:________ Date: ___/___/___

PS11 Journal Sheet #3

FORCE, WORK, AND SIMPLE MACHINES

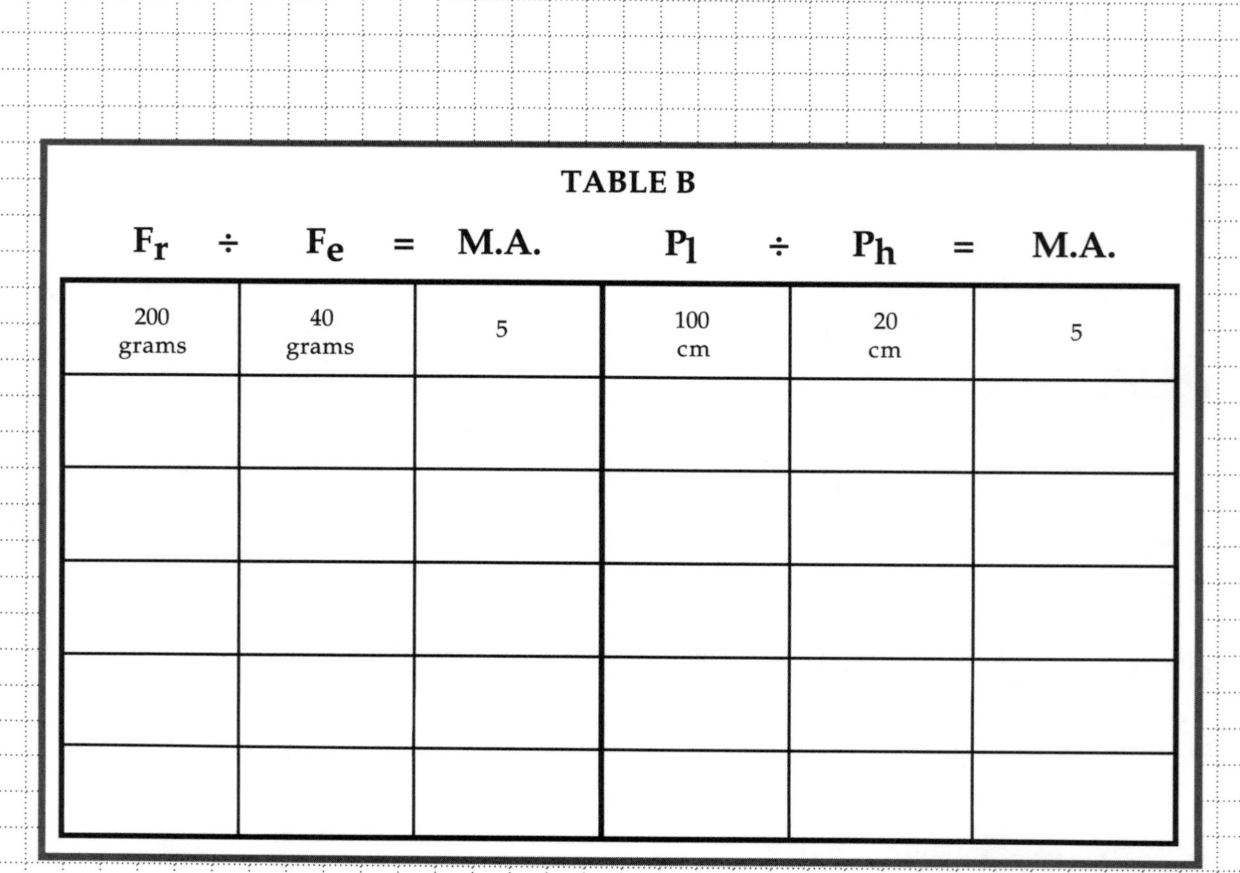

TABLE B

F_r ÷	F_e =	M.A.	P_l ÷	P_h =	M.A.
200 grams	40 grams	5	100 cm	20 cm	5

<u>NOTE</u>: In all examples you can ignore the acceleration of gravity since it is the same for all objects. In the first example, a **200-gram** car (**F_r**) is balanced by a **40-gram** mass (**F_e**) on a ramp with a length of **100 centimeters** (**P_l**) and a height of **20 centimeters** (**P_h**) above the table. In both calculations the **mechanical advantage** of the ramp (**M.A.**) is equal to **5**.

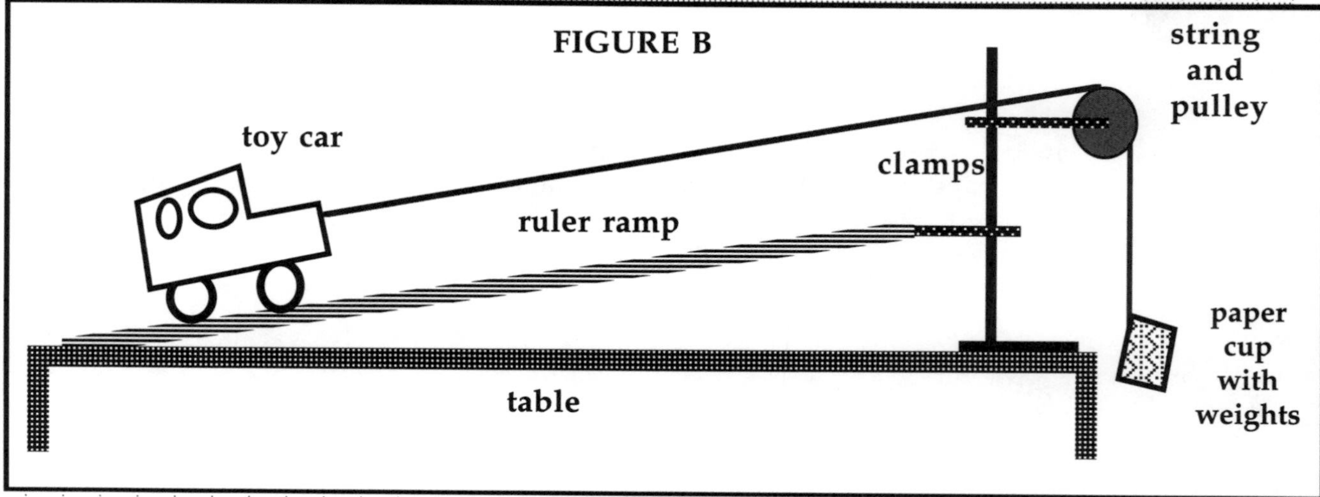

FIGURE B

PS11 **Lesson #4**

FORCE, WORK, AND SIMPLE MACHINES

Work Date: ____/____/____

LESSON OBJECTIVE

Students will demonstrate how a pulley produces a mechanical advantage.

Classroom Activities

On Your Mark!

Draw the **compound pulley** shown in Illustration A in the Teacher's Classwork Agenda and Content Notes. Explain that this type of pulley described by **Hero of Alexandria** around 100 A.D. produces a mechanical advantage equal to 2; because there are two cables supporting the resistance force or load. Ask students to consider the following situation: You and two friends are out hiking. One of your friends slips and falls over a ledge. He is hanging on for dear life. You grab hold of your friend's arm and try to lift his 100 pounds. You do not have the strength. Luckily, your second friend grabs the first friend's other arm and, together, you both lift your endangered pal to safety. How much weight did each of you have to lift? Answer: 50 pounds. Why? Because you each lifted half the weight. A pulley acts under the same principle. Each string or cable of a pulley carries only a portion of the load's weight. The mechanical advantage of a pulley is equal to the number of cables supporting the load.

Get Set!

Have the set-up shown in Figure C on Journal Sheet #4 on display before students enter class.

Go!

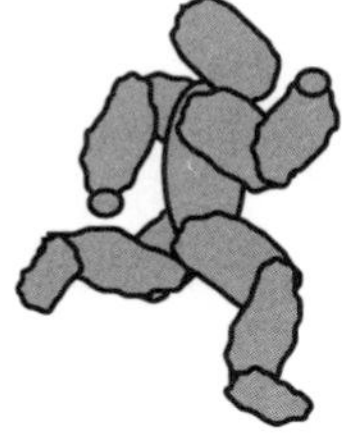

Challenge students to construct the pulley system shown in Figure C on Journal Sheet #4. Have them use a variety of small and large brass weights to demonstrate that their pulley has a mechanical advantage of 5. The construction of this device is not as easy as it looks. It requires a good deal of manual dexterity and may prove frustrating for many students. Tell students to keep a positive attitude and work slowly and cooperatively to achieve success.

Materials

single, double, and triple pulleys, brass weights (or pennies at 3 grams/penny), string, ring stand, clamps, paper cups

Name: ______________________ Period:________ Date: ____/____/____

PS11 Journal Sheet #4

FORCE, WORK, AND SIMPLE MACHINES

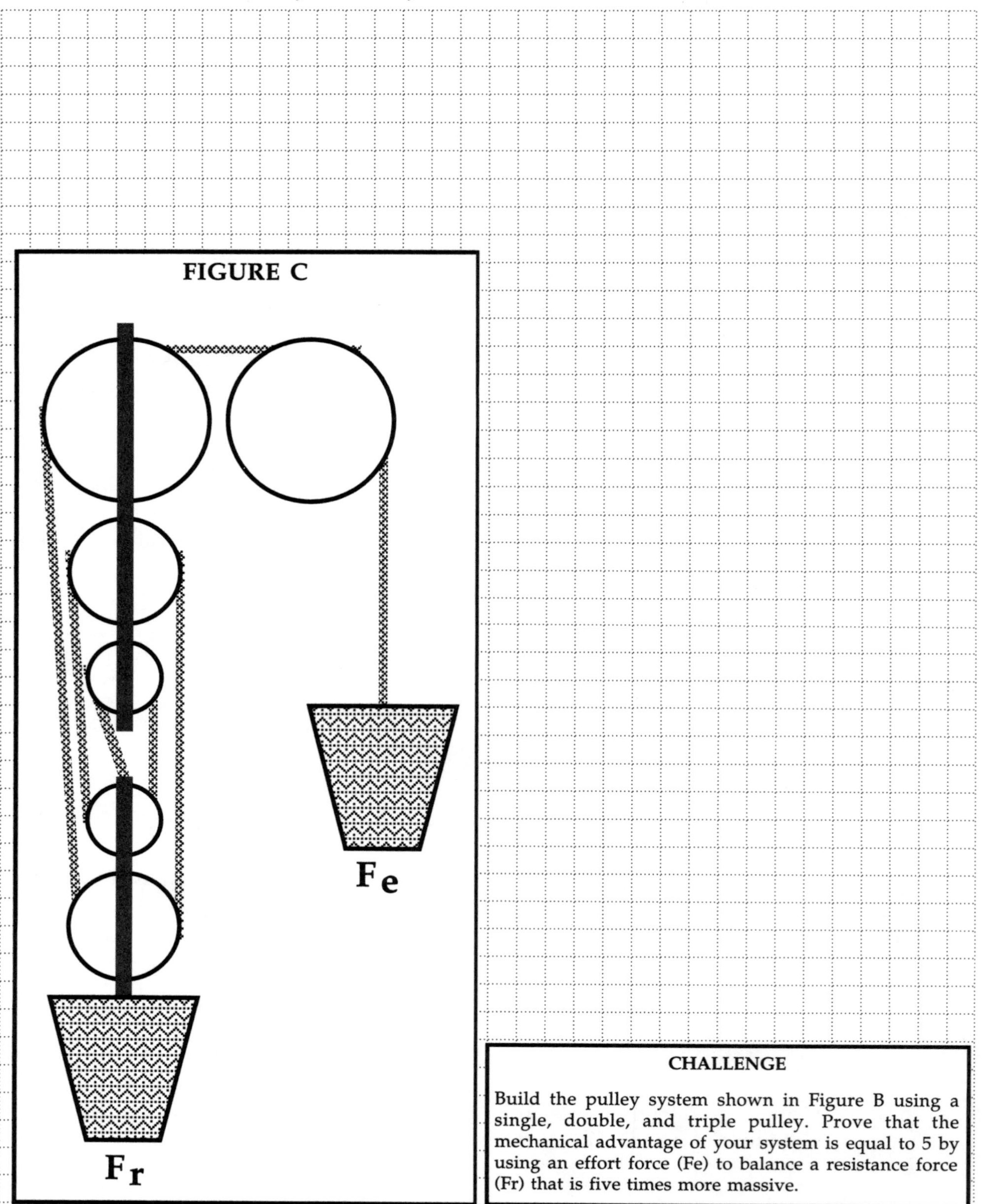

CHALLENGE

Build the pulley system shown in Figure B using a single, double, and triple pulley. Prove that the mechanical advantage of your system is equal to 5 by using an effort force (Fe) to balance a resistance force (Fr) that is five times more massive.

Name: ______________________________ **Period:**________ **Date:** ____/____/____

PS11 Review Quiz

Directions: Keep your eyes on your own work.
Read all directions and questions carefully.
THINK BEFORE YOU ANSWER!
Watch your spelling, be neat, and do the best you can.

CLASSWORK	(~40):	_______
HOMEWORK	(~20):	_______
CURRENT EVENT	(~10):	_______
TEST	(~30):	_______
TOTAL	(~100):	_______

(A ≥ 90, B ≥ 80, C ≥ 70, D ≥ 60, F < 60)

LETTER GRADE: _______

TEACHER'S COMMENTS: ____

____

FORCE, WORK, AND SIMPLE MACHINES

TRUE–FALSE FILL-IN: If the statement is true, write the word TRUE. If the statement is false, change the underlined word to make the statement true. *10 points*

_______________ 1. A person trying but failing to lift a heavy object <u>has not</u> done any work.

_______________ 2. Multiplying force times distance gives a measure of <u>power</u>.

_______________ 3. In the Metric System, a unit of measure for <u>force</u> is the newton-meter or joule.

_______________ 4. In the <u>English</u> System, a unit of measure for work is the foot-pound.

_______________ 5. A person or machine applying a 6-newton force over a distance of 3 meters has done <u>9 joules</u> of work.

_______________ 6. Machines <u>always</u> save work.

_______________ 7. The mechanical advantage produced by a machine can be calculated by <u>dividing</u> the resistance force by the effort force.

_______________ 8. A(n) <u>lever</u> is a ramp or tilted surface.

_______________ 9. Increasing the height of an inclined plane <u>decreases</u> the mechanical advantage of the plane.

_______________ 10. Increasing the distance of the effort force from the fulcrum <u>decreases</u> the mechanical advantage of the lever.

List four (4) things that a machine can do. *4 points*

(A) ____

(B) ____

(C) ____

(D) ____

PROBLEM

Directions: Read the problem carefully. Show all formulas and mathematical calculations used in solving the problem. Be sure to include the correct unit of measure with your answer in the space marked "Final Answer." *6 points*

A group of ants does 16,000 ergs of work moving a pebble of sand blocking the opening of their ant hill. The mass of the pebble is 4 grams. The ants are near the surface of the earth where gravity accelerates falling objects at the rate of about 1,000 centimeters per second per second. How far did the ants move the pebble?

Final Answer: ______________

______________________________ Student's Signature

______________________________ Parent's Signature

____/____/____ Date

PS12

MACHINE EFFICIENCY AND POWER

TEACHER'S CLASSWORK AGENDA AND CONTENT NOTES

Classwork Agenda for the Week

1. Students will explain the relationship between "work" and "power."
2. Students will draw a pulley system with a mechanical advantage of 50 and construct it to demonstrate that no machine is ever 100% efficient.
3. Students will show how a wheel and axle assembly produces a mechanical advantage and how gears work according to the same principle.
4. Students will calculate the mechanical advantage of a bicycle and discuss why the bicycle is one of the most efficient machines.

Content Notes for Lecture and Discussion

The concepts of **force** and **power** are frequently used interchangebly in everyday language; but to a scientist the two ideas are very different. As previously defined, force is a "push or pull" on an object and related to the mass of the object and the degree to which it is being accelerated. According to Newton's Second Law of Motion, **f = m × a**. **Work** is the application of a force through a distance: **W = F × D**. **Power**, on the other hand, is the amount of work done by a worker or machine in a given amount of time. Power is work divided by time: **P = W ÷ t**. In the Metric System (MKS) power is measured in **joules per second** or **watts** after the inventor of the steam engine: Scottish engineer **James Watt** (b. 1736; d. 1819). In the English System, the unit of measure for power is **horsepower**. A machine has 1 horsepower when it can move a 55-pound load a distance of 1 foot in 1 second.

In Lesson #1 students will reinforce their understanding of the meaning of "work" and relate it to the concept of "power."

Ever since people have used machines they have been aware that machines get warm while in use. **Friction** between the parts of machines produces heat which is radiated away from the machine as wasted energy. Instead of the worker's energy being put to productive use it is lost to the universe forever. The hotter a machine gets while in use, the less efficient it is. A measure of the amount of work a machine really does compared to the amount of work it might ideally do if there were no frictional forces is a measure of the **efficiency** of the machine. **Machine efficiency** is measured using the following formula: **M.E. =** $(\mathbf{W_o \div W_i}) \times \mathbf{100}$, where **M.E.** is machine efficiency, $\mathbf{W_o}$ is work output and $\mathbf{W_i}$ is work input. The ratio between these two quantities is multiplied by **100** to give a **percentage**.

In Lesson #2, students will review and discuss why machines cannot save work and learn to calculate the efficiency of a ramp and pulley system that they build themselves.

Since ancient times, engineers have searched for better ways to ease the movement of wheels. **Gears** and **bearings** are the primary tools used to accomplish this. By using gears, engineers discovered that they could increase force and speed depending upon how gears of varying sizes were connected.

In Lesson #3 students will examine the mechanical advantage produced by a **wheel and axle** assembly. In a wheel and axle assembly, a large wheel is used to drive a small wheel or vice versa. Gears function by the same principle as the wheel and axle.

PS12 Content Notes *(cont'd)*

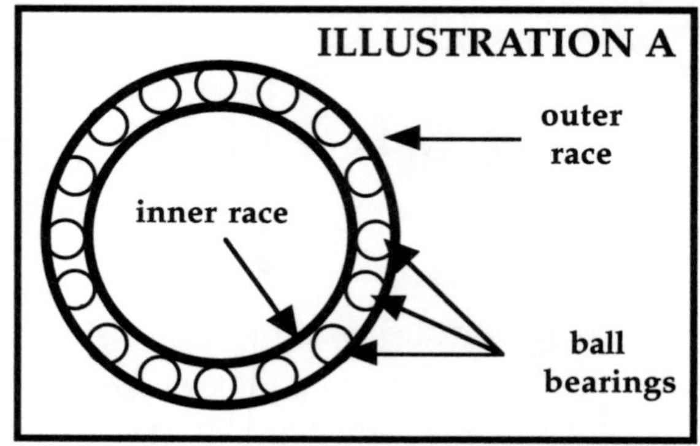

Bearings and **gears** have been around for about 3,000 years. They were first used in Europe to facilitate the movement of wheels on horse-drawn carts and wagons. Modern **ball bearings** like the one pictured in Illustration A reduce friction by encasing steel balls between two steel rings. One moving ring or "race" is fixed to the moving part of the machine while the other race is attached to the stationary part of the machine. The ball bearings encased between the two races roll against one another to reduce friction.

In Lesson #4 students will examine the gears and chains on the bicycles of fellow students. They will determine the mechanical advantage produced by the gears. They may be interested to discover that the bicycle is a relatively recent invention. The first bicycle was invented in Great Britain in 1839 by a Scottish blacksmith named **Kirkpatrick Macmillan** (b. 1813; d. 1878). Macmillan's bicycle used rods and cranks to increase the mechanical advantage of the peddling rider. His design was improved by British inventer, **James Starley** (b. 1830; d. 1881) in 1870. Starley's bicycle was called a "Penny Farthing" or "high-wheeled bicycle" because the front wheel was two to three times larger than the back wheel. His nephew, John Starley, designed the modern bicycle in 1885 with tires of equal sizes and gears attached by chains.

ANSWERS TO THE HOMEWORK PROBLEMS

1. M.E. = $(W_o \div W_i) \times 100$
   = $(40_j \div 50_j) \times 100$
   = 0.8×100
   = 80%
2. P = W ÷ t
   = F × D ÷ t
   = 5,500 lbs × 100 feet ÷ 100 seconds
   = 5,500 ft.-lbs. per second
   Since 1 horsepower is 55 ft.-lbs. of work in 1 second, then . . .
   = 100 horsepower.
3. The M.A. of the large gear over the small gear is 48 ÷ 16 = 3. The circumference of the rear wheel is $C = 2\pi r = 2(3.14)40\text{cm} = 251.2$ cm. With one revolution of the pedal the rear wheel will travel three revolutions at 251.2 cm/revolution. *Answer:* The bike will travel 753.6 cm.

ANSWERS TO THE END-OF-THE-WEEK REVIEW QUIZ

1. C
2. B
3. A
4. A
5. E
6. heat or wasted energy
7. increase
8. true
9. power
10. horsepower or ft.-lb./s
11. 12 divisions ÷ 4 divisions = 3
12. 5 cables supporting the load
13. 165 ÷ 15 = 11 lbs.
14. The work output is 1,650 ft.-lbs. Since the work input is 3,300 ft.-lbs., the machine's efficiency is 50%
15. 3,300 ft.-lbs.÷3 seconds is 1,100 ft.-lbs./s or 20 horsepower

Name: ______________________ **Period:**________ **Date:** ___/___/___

PS12 Fact Sheet

MACHINE EFFICIENCY AND POWER

CLASSWORK AGENDA FOR THE WEEK

(1) Show the relationship between work and power.
(2) Build a pulley system with a mechanical advantage of 50 to demonstrate that machines are never 100% efficient.
(3) Examine the effectiveness of gears in producing mechanical advantage.
(4) Calculate the mechanical advantage of a bicycle and discuss why it is one of the most efficient machines.

All machines have moving parts. Moving parts make contact with one another to cause **friction** that produces **heat.** Heat is wasted energy that decreases the amount of work done by a machine. The less friction there is between the moving parts of a machine the more efficient it is. That is why mechanics lubricate machines with grease or oil. Lubricants reduce friction and improve the efficiency of machines.

Machine efficiency is a fraction (or percentage) of the total amount of work a machine could do if there were no friction. Machine efficiency, or **M.E.**, can be calculated using the following formula:

$$\mathbf{M.E. = (W_o \div W_i) \times 100\%}$$

In the above formula, W_o is **work output** or the amount of work the machine does. W_i is the **work input** of the machine or the amount of effort that the worker puts into the machine to get it to work (i.e., fuel). The efficiency of a machine would be 100% if the "input" work were equal to the "output" work. Since friction always decreases the output work done by a machine, no machine can ever be 100% efficient. The efficiency of all machines is always less than 100%.

A machine's **power** is the amount of work it does in a given amount of time. Power **(P)** can be calculated using the following formula:

$$\mathbf{P = W \div t}$$

In this formula, **W** is work and **t** is time. Since the unit of measure for work is the **joule** in the Metric System, the unit of measure for power is the **joule per second**—also called a **watt**. In the English System of measurement, where work is measured in foot-pounds, the unit of measure for power is the **foot-pound per second**. If a machine can do **55 foot-pounds of work in 1 second** it has **1 horsepower**.

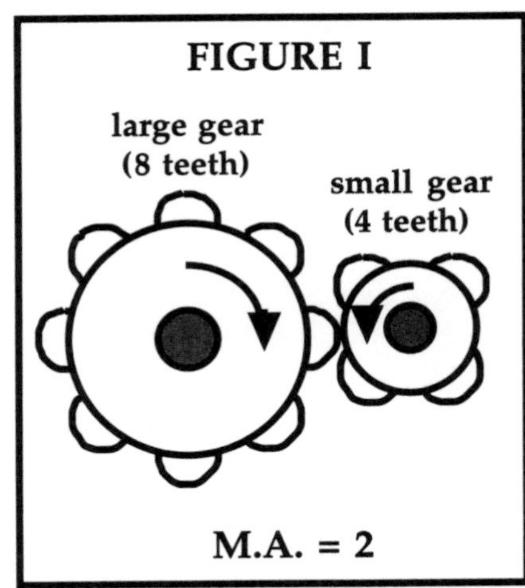

A bicycle is of one the most efficient machines there is. On a bicycle, the rider's effort force and input work are transferred using lubricated gears and chains that greatly reduce friction. Unlike the gears shown in Figure I, gears whose teeth are directly interlocked, the teeth on bicycle gears are interlocked by the chain. A bicycle chain causes both gears to turn in the same direction. Notice that the gears in Figure I turn in the opposite direction. The number of teeth on interlocked gears determines how much force and speed will be increased.

Homework Directions

SHOW ALL MATHEMATICAL FORMULAS AND CALCULATIONS IN SOLVING PROBLEMS #1, #2, AND #3. BE SURE TO INCLUDE CORRECT UNITS OF MEASURE WITH YOUR ANSWER.

1. What is the efficiency of a machine that does 40 joules of work if the fuel burned by the machine contains 50 joules of energy? How many joules of energy were wasted as heat due to friction?
2. A truck moves 5,500 pounds of load 100 feet in 100 seconds. How much power does the truck have?
3. The large pedal gear on a bicycle has 48 teeth. One of the smaller gears connected to the rear wheel by a chain has 16 teeth. The radius of the pedal is 20 centimeters and the radius of the rear wheel is 40 centimeters. How far will the bicycle move with one turn of the pedal?

Assignment due: __________

______________________________ Student's Signature

______________________________ Parent's Signature

____/____/____ Date

MACHINE EFFICIENCY AND POWER

Work Date: ___/___/___

LESSON OBJECTIVE

Students will explain the relationship between "work" and "power."

Classroom Activities

On Your Mark!

Begin with a review of the concepts of **force** and **power.** Remind students that a force is a push or pull on an object and that, according to Newton's Second Law of Motion, force depends on mass and acceleration: $\mathbf{f = m \times a}$. Have them recall that work is the application of a force through a distance: $\mathbf{W = F \times D}$. Explain that power is the amount of work done in a given amount of time. Write the formula for calculating the power of a machine, $\mathbf{P = W \div t}$, and explain each variable. Have students copy the formula on Journal Sheet #1. Draw a picture like the one shown in Illustration A to help you explain the meaning of horsepower. End your lecture with a discussion of the concept of **machine efficiency**. Ask students this question: "Is an automobile an efficient machine?" Answer: No. An automobile burns four to five times as much gasoline to produce heat as it does to do keep the car moving. A car, like all other combustion engines, is about 20% efficient. Compared to a bicycle, a car is extremely inefficient! Write the formula for calculating the efficiency of a machine on the board, $\mathbf{M.E. = (W_o \div W_i) \times 100}$, and have students copy the formula on Journal Sheet #1.

Get Set!

Show students how to calculate the **power** of machines making sure they remember to use the correct units of measure. Show students how to calculate the **efficiency** of machines, making sure they use the correct units of measure. Use the examples on Journal Sheet #1 to aid you in creating additional examples.

Go!

After making sure that students feel comfortable with calculating power and machine efficiency in your mathematical examples, have them complete the Practice Problems on Journal Sheet #1. Review the problems to make sure they understand the algorithms and units of measure.

Materials

Journal Sheet #1

ANSWERS TO THE PROBLEMS ON JOURNAL SHEET #1

1. M.E. $= (W_o \div W_i) \times 100$ <u>and</u> $W = F \times D$; so . . .
   $= [(250n \times 2m) \div (100n \times 10m)] \times 100$
   $= [500j \div 1{,}000j] \times 100 = 50\%$
2. P $= W \div t = (1{,}000n \times 10m) \div 2s = 5{,}000$ watts
3. P $= W \div t = (110 \text{ lbs.} \times 50 \text{ ft}) \div 5s$
   $= 5{,}500$ ft.-lbs. $\div$ 5s = 1,100 ft.-lbs./s or 20 horsepower

Name: ________________________________ **Period:**________ **Date:** ___/___/___

PS12 Journal Sheet #1

MACHINE EFFICIENCY AND POWER

PRACTICE PROBLEMS

Note: Show all formulas, calculations, and correct units of measure for each problem.

1. A worker uses a crank to wind up 10 meters of pulley cable with 100 newtons of force. He is able to lift a 250-newton load a distance of 2 meters off the ground. What is the efficiency of his machine? How many joules of energy were wasted as a result of friction in his machine?

2. What is the power of a machine able to move a 1,000-newton load a distance of 10 meters in 2 seconds?

3. What is the power of a machine able to move a 110-pound load 50 feet in 5 seconds?

MACHINE EFFICIENCY AND POWER

Work Date: ____/____/____

LESSON OBJECTIVE

Students will draw a pulley system with a mechanical advantage of 50 and construct it to demonstrate that no machine is ever 100% efficient.

Classroom Activities

On Your Mark!

Review students' calculations in Lesson #1. Draw Illustration B on the board. Explain that a ramp that is 100 centimeters in length and 10 centimeters in height will have a mechanical advantage of 10. Combining the ramp with a pulley that has 5 cables supporting the load will increase the mechanical advantage of the machine to 50. After discussing these points with the class, pose the following situation: If their toy car weighs 200 grams, the mass in the cup needed to balance the car on the ramp should be a little more than 4 grams. The extra mass is needed to overcome the friction between the machine parts.

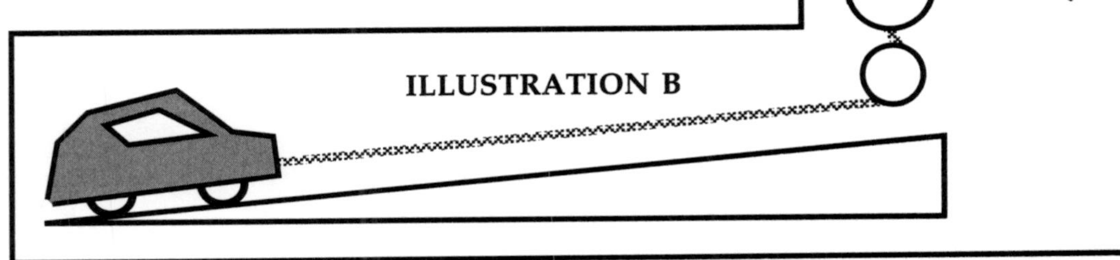

ILLUSTRATION B

Get Set!

Have students copy Illustration B on Journal Sheet #2 and record the measurements of the ramp (i.e., length and height) and the mechanical advantage of the pulley. Have them show their calculations for determining the M.A. of the entire machine.

Go!

Give students ample time to construct their pulley system and balance their toy car on the ramp. Conclude the lesson by discussing the amount of mass necessary to support the load. Was it greater than expected? Answer: Yes. Why? Due to friction.

Materials

single, double, and triple pulleys; brass weights (or pennies at 3 grams/penny); string; ring stand; clamps; paper cups; toy cars or trucks

Name: ______________________________ Period: ________ Date: ____/____/____

PS12 Journal Sheet #2

MACHINE EFFICIENCY AND POWER

Directions: Draw a ramp and pulley system made of metric rulers, a ring stand and clamps, and pulleys and string. Your machine should allow you to raise a toy car or truck up your ramp with a mechanical advantage of 50.

PS12 Lesson #3

MACHINE EFFICIENCY AND POWER

Work Date: ___/___/___

LESSON OBJECTIVE

Students will show how a wheel and axle assembly produces a mechanical advantage and how gears work according to the same principle.

Classroom Activities

On Your Mark!

Use the metal bar from a ring stand to perform the following demonstration: (1) Ask for a "strong" student volunteer. (2) Instruct the volunteer to grasp the metal bar with both hands pressed together at the center of the bar as shown in Illustration C. (3) Place your hands at the ends of the bar and challenge the student to prevent you from rotating the bar in a clockwise direction. (4) When the student is ready, gently rotate the bar against their opposing force. This will be extremely easy, since your hands are placed at the ends of the bar, you have a mechanical advantage over the student. (5) Choose a "weak" student to put his or her hands at the end of the bar so that you can try to prevent them from rotating the bar. You will find it extremely difficult to stop them because they now have the advantage. Prompt students to recall

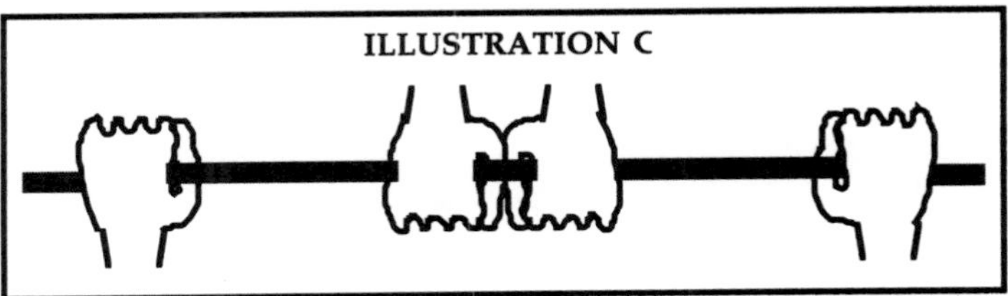
ILLUSTRATION C

the principle of a lever: The farther away from the fulcrum one applies force, the easier it is to move a mass close to the fulcrum. Explain that a **wheel and axle** assembly works on the same principle. Explain that **gears**, also, work according to this idea.

Get Set!

Draw the wheel and axle assembly shown in Illustration D and label the parts. Students can copy this information on Journal Sheet #3. Give them the formula **M.A. = $R_w \div R_a$** with which to calculate the mechanical advantage of a wheel and axle assembly. Explain that gears work by the same principle although counting and comparing the number of teeth on interlocking gears is easier than measuring the radii of the wheel and axle.

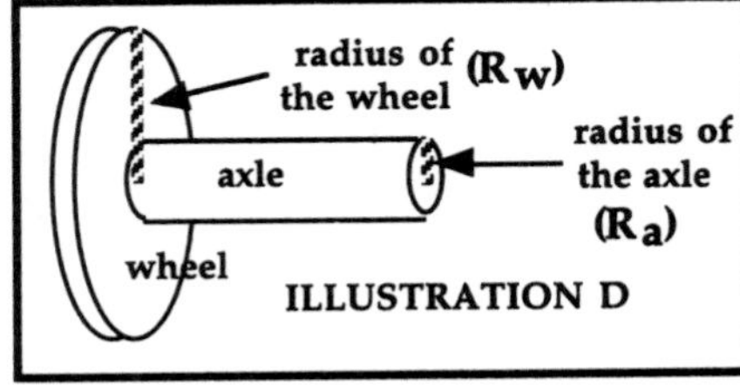

ILLUSTRATION D

Go!

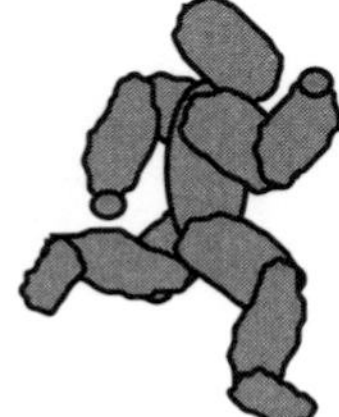

Instruct students to examine the relationship between gears A, B, and C drawn in Figure A on Journal Sheet #3. Have them discuss and record the mechanical advantages of each interlocking pair of gears. Ask them to show the directions in which the gears will turn when in motion.

Materials

metal bar from a ring stand, gears for show (if available), Journal Sheet #3

Name: ______________________________ Period:________ Date: ___/___/___

PS12 Journal Sheet #3

MACHINE EFFICIENCY AND POWER

Directions: Label the parts of this diagram as shown by your instructor. Copy the formulas used to calculate the mechanical advantage of a wheel and axle.

FIGURE A

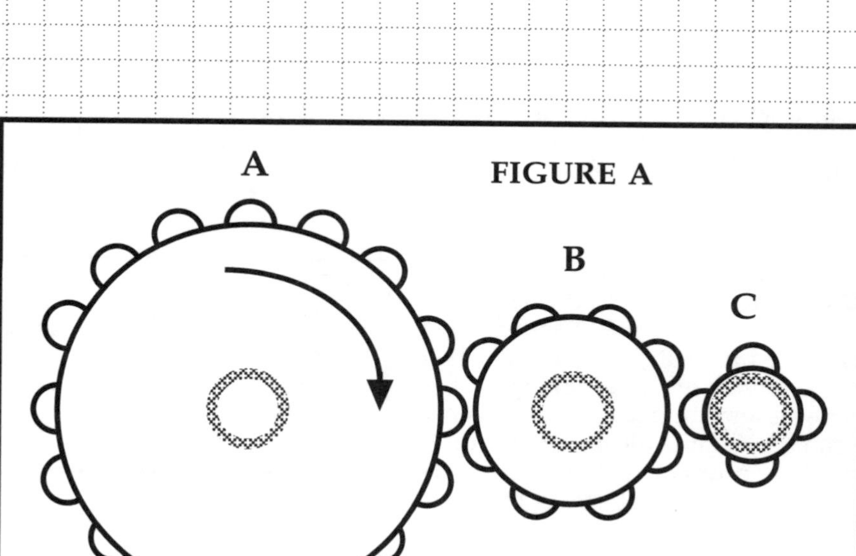

Examine the relationship between gears A, B, and C shown above. Discuss with your classmates the mechanical advantage produced by using the large gear to drive the small gear. Discuss the direction in which the gears will turn. Write down the important points of your discussion.

PS12 **Lesson #4**

MACHINE EFFICIENCY AND POWER

Work Date: ____/____/____

LESSON OBJECTIVE

Students will calculate the mechanical advantage of a bicycle and discuss why the bicycle is one of the most efficient machines.

Classroom Activities

On Your Mark!

Arrange with several students and their parents/guardians to have bicycles brought to school for the purpose of this lesson. Ordinary bikes with few speeds are preferable to bikes with multiple gears. The latter are more difficult to examine. Review the history of the modern bicycle using the information in the Teacher's Agenda and Content Notes. Ask students to read Questions #1, #2, and #3 on Journal Sheet #4.

Get Set!

Have students assist you in placing the bicycles upside down, resting on seat and handle bars, on the floor or sturdy tables. Explain the Safety Precautions that must be observed when examining the bicycles (i.e., "Do not spin the tires!" "Do not put your fingers near the chain or gear teeth!" etc.). Show them how to measure the radii of the wheels and gears using a metric ruler.

Go!

Instruct students to complete the assignment on Journal Sheet #4.

Materials

bicycles, metric rulers

Name: ______________________________ Period:________ Date: ____/____/____

PS12 Journal Sheet #4

MACHINE EFFICIENCY AND POWER

radius of the foot pedal gear	radius of the rear wheel gear	M.A. of the gear combination	teeth on the foot pedal gear	teeth on the rear wheel gear	M.A. of the gear combination

<u>Note:</u> Show all formulas and calculations on this journal sheet in order to answer these questions about the bicycle you examine.

1. How many times will the rear bicycle wheel turn for every single turn of the foot pedal? __________

2. How far will the bicycle move with a single turn of the foot pedal? (Note: The circumference of a circle can be calculated using the formula **$C=2\pi r$**, where **C** is circumference, π **= 3.14**, and **r** is the radius of the circle) __________

3. Why is it necessary to lubricate gears and chains on a bicycle?__________

____

____

Name: ______________________________ **Period:**________ **Date:** ____/____/____

PS12 Review Quiz

Directions: Keep your eyes on your own work.
Read all directions and questions carefully.
THINK BEFORE YOU ANSWER!
Watch your spelling, be neat, and do the best you can.

CLASSWORK	(~40):	_______
HOMEWORK	(~20):	_______
CURRENT EVENT	(~10):	_______
TEST	(~30):	_______
TOTAL	(~100):	_______

(A ≥ 90, B ≥ 80, C ≥ 70, D ≥ 60, F < 60)

LETTER GRADE: _______

TEACHER'S COMMENTS: ____

____

MACHINE EFFICIENCY AND POWER

MULTIPLE CHOICE: Choose the letter of the word or phrase that best answers the question or completes the statement. *10 points*

______ 1. A machine's ability to multiply the force of a worker is a measure of its __________.
(A) friction (C) mechanical advantage (E) usefulness
(B) machine efficiency (D) power

______ 2. The amount of work wasted by a machine can be found by calculating its __________ .
(A) friction (C) power (E) temperature
(B) efficiency (D) advantage

______ 3. Which is a unit of measure for power?
(A) watt (C) newton (E) joule
(B) erg (D) dyne

______ 4. Which of the following *is not* a kind of lever.
(A) an axe blade (C) a crank (E) a pulley
(B) a gear (D) a crowbar

______ 5. Worker X moves a 2-pound box a distance of 12 feet. Worker Y moves a 6-pound box a distance of 4 feet. Worker Z moves a 3-pound box a distance of 8 feet. Which worker does the most work?
(A) X (C) Z (E) all do the same amount of work
(B) Y (D) X = Z

TRUE–FALSE FILL-IN: If the statement is true, write the word TRUE. If the statement is false, change the underlined word to make the statement true. *10 points*

______________ 6. The moving parts of a machine cause friction which produces a mechanical advantage.

______________ 7. Machines use grease or oil to decrease machine efficiency.

______________ 8. A machine cannot be 100% efficient.

_______________ 9. A machine's mechanical advantage is the amount of work it does in a given amount of time.

_______________ 10. In the English System of measurement, where work is measured in foot-pounds, the unit of measure for power is the joule.

PROBLEM

Directions: Study the machine shown below and answer questions #11 through #15.

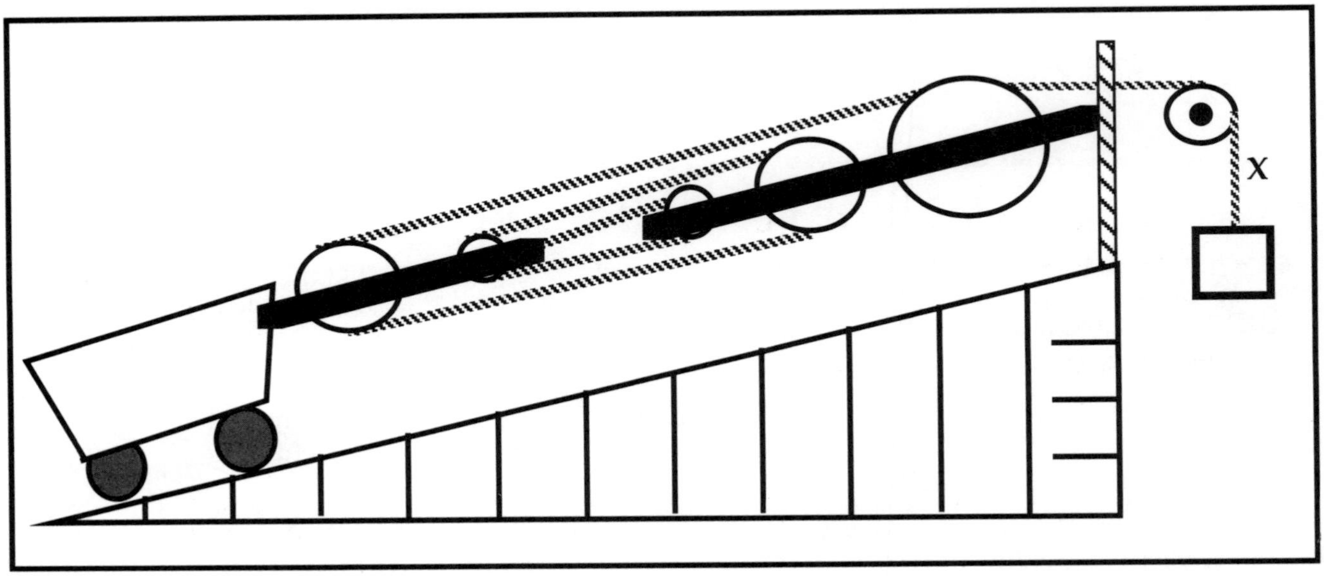

11. What is the mechanical advantage of the inclined plane shown in the figure above? _______________
12. What is the mechanical advantage of the pulley shown in the figure above? _______________
13. What is the least amount of effort force that should be required to lift a 165-pound load by pulling on rope X? _______________
14. If the amount of work put into lifting the 165-pound load 10 feet up the inclined plane is 3,300 foot-pounds, what is the efficiency of the machine? _______________
15. If the time it takes to lift the load up the inclined plane is 3 seconds, what is the power of this machine? _______________

_______________ Student's Signature

_______________ Parent's Signature

___/___/___ Date

PS13

ENERGY

TEACHER'S CLASSWORK AGENDA AND CONTENT NOTES

Classwork Agenda for the Week

1. Students will set up a pendulum and discuss how potential and kinetic energy can be transformed from one to the other.
2. Students will construct a model of a steam turbine.
3. Students will explain the purpose of each stroke of the standard 4-stroke internal combustion engine used in most modern automobiles.
4. Students will pick out pertinent information from an electric bill.

Content Notes for Lecture and Discussion

The concept of **energy** has evolved over the centuries as scientists increased their store of knowledge about people's ability to do work. The development of the idea of energy began with the ancient Greeks' study of **mechanics** in their quest to understand why and how objects moved. The concepts of **potential** and **kinetic energy** have their origins in this tradition. The mathematician-philosopher and contemporary of Sir Isaac Newton, **Gottfried Wilhelm Liebniz** (b. 1646; d. 1716) was among the first to suggest that "energy of composition or position (a.k.a., potential energy)" and "energy of motion (a.k.a., kinetic energy)" were interchangeable. Liebniz pointed out that either form of energy could be transformed to the other. This idea led to the **First Law of Thermodynamics: The Law of Conservation of Energy**. Briefly summarized, the law states that energy can neither be created nor destroyed. Of course all matter has both potential and kinetic energy because no piece of matter in the universe is ever at rest. The total energy of an object is the sum of its potential and kinetic energy.

In the eighteenth and nineteenth centuries, the careful study of **heat, light, magnetism**, and **electricity** rooted the idea of energy in the idea of **forces**. The **invention of the steam engine** by British engineer **Thomas Savery** (b. 1650; d. 1715) in 1698 to pump water from flooded mines, a design perfected by **James Watt** (b. 1736; d. 1819) in 1782, led to the **Second Law of Thermodynamics: The Law of Entropy**. Briefly stated, the Second Law declares that energy is constantly being changed from useful and organized forms to less useful and more disorganized forms. The **force of friction** is largely responsible for this phenomenon. It was **William Thomson Kelvin** (b. 1824; d. 1907) who formalized the Laws of Thermodynamics in concrete mathematical form. Students are most familiar with the idea of heat energy. Use their notions of heat energy to discuss how heat is transfered by **conduction** (transfer through solids), **convection** (transfer through fluids) and **radiation** (transfer through solids, fluids, and a vacuum by electromagnetic forces).

In Lesson #1 students will use a pendulum to demonstrate the difference between potential and kinetic energy. They will repeat an experiment performed by **Galileo Galilei** (b. 1564; d. 1642) in which he discovered that the length, and not the mass, of a pendulum determines the period of the pendulum's swing. In the process, students will see how potential energy is constantly being changed to kinetic energy, and vice versa, during the pendulum's motion. They will also conclude that the pendulum loses energy to its environment as the result of its contact with the air (i.e., **air friction**).

In Lesson #2 students will construct a model of a steam turbine and explain how it works. This lesson includes a brief discussion of how electricity is generated.

In Lesson #3 students will examine the workings of the 4-stroke internal combustion engine used in most modern automobiles.

PS13 Content Notes *(cont'd)*

In Lesson #4 students will bring their ideas about energy "down to earth" by examining a bill from the electric company. This activity will reinforce the notion that "machines can never save energy." If machines did save energy, there would be a surplus, an oversupply, that could be given back to consumers free of charge. This lesson will impress students with the importance of conserving our energy resources.

ANSWERS TO THE HOMEWORK PROBLEMS

1. Answers will vary in this assignment.
2. Work and energy are based on the same units of measure as shown in the following equations. You may wish to reserve this problem for high achieving or gifted students.

$$\mathbf{W} = F \times D = m \times c^2 = \mathbf{E}$$
$$= m \times a \times D = m \times c \times c =$$
$$\textbf{joules} = \text{kilograms} \times \frac{\text{meters}}{\text{sec}^2} \times \text{meters} = \text{kilograms} \times \frac{\text{meters}}{\text{sec}} \times \frac{\text{meters}}{\text{sec}} = \textbf{joules}$$

ANSWERS TO THE END-OF-THE-WEEK REVIEW QUIZ

1. E
2. A
3. D
4. E
5. E
6. do work
7. true
8. true
9. Albert Einstein
10. can

FIRST PROBLEM

$$\text{G.P.E} = mgh = 5 \text{ kg} \times 9.8 \text{ m/s}^2 \times 4 \text{ m} = 196 \text{ joules}$$

SECOND PROBLEM

$$\text{K.E.} = mv^2 \div 2 = [(10 \times 10^3 \text{ kg}) \times (2 \times 10^3 \text{ m/s}) \times (2 \times 10^3 \text{ m/s})] \div 2$$
$$= 40 \times 10^9 \text{ kg-m}^2/\text{s}^2 \div 2 = 20 \times 10^9 \text{ joules}$$

Name: ______________________ Period:________ Date: ___/___/___

PS13 Fact Sheet

ENERGY

CLASSWORK AGENDA FOR THE WEEK

(1) Use a pendulum to demonstrate the difference between potential and kinetic energy.
(2) Construct a model of a steam turbine and explain how it works.
(3) Label a diagram of a 4-stroke automobile engine and explain how it works.
(4) Examine a bill from the electric company.

Energy is the ability to do work. Like work, energy is measured in joules after the English physicist **James Prescott Joule** (b. 1818; d. 1889). The amount of energy an object has depends on its **composition** (the kind of matter it contains), its **position** (where the matter is located), and its **velocity** (how fast the matter is moving).

Potential energy describes the energy that an object has because of its composition or position. A stick of dynamite is just a package of dangerous chemicals . . . until it explodes! Dynamite was invented by Swedish chemist **Alfred Nobel** (b. 1833; d. 1896) in 1867. Dynamite contains **chemical potential energy** that can be released under certain conditions with explosive results. All of the atoms that make up matter have **nuclear potential energy**. The famous American scientist, **Albert Einstein** (b. 1879; d. 1955) discovered that small amounts of matter can release huge amounts of energy. The amount of energy released by matter during a **nuclear reaction** can be calculated using Einstein's famous formula: $\mathbf{E=mc^2}$. In this familiar equation, **E** is energy, **m** is the mass of the matter, and **c** is the velocity of light (300,000 kilometers per second). All matter has **gravitational potential energy** as well; because all matter in the universe is influenced by the force of gravity. A large boulder balanced on the edge of a cliff is just an interesting natural sculpture . . . until it falls!

The gravitational potential energy of an object can be calculated using the following formula:

$$\text{G.P.E.} = mgh$$

where **G.P.E.** is gravitational potential energy, **m** is mass, **g** is the acceleration of gravity (9.8 meters per second per second), and **h** is the height of the small object (like a boulder) above a much larger object (like the earth). Remember that variables—like **m**, **g**, and **h**—placed next to one another in an equation are multiplied when solving problems using that equation.

Kinetic energy is energy of motion. Any moving object has the ability to do work. The faster an object is moving, or the more massive it is, the more work it will do when it collides with another object. The kinetic energy of an object can be calculated using the following formula:

$$\text{K.E.} = mv^2 \div 2$$

where **K.E.** is kinetic energy, **m** is the mass of the moving object and **v** is the velocity of the moving object.

Energy comes in a variety of forms that can be changed from one form to another: **heat energy**, **chemical energy**, **nuclear energy**, **electrical energy**, **solar energy**, **mechanical energy**, etc. Electric generators take advantage of the fact that kinetic energy can be transformed into electrical energy. The motion of water falling over a dam, or steam escaping from a kettle (i.e., kinetic energy), can be used to spin a magnet inside a coil of wire. The magnet causes electrons in the metal wire to move. Moving electrons produce an **electric current**. A variety of **fuels** (i.e., the chemical energy stored in coal or oil) are used to spin the magnet inside the coil of wire to produce electric current.

Homework Directions

1. Make a chart of ten (10) energy transformations that occur in your home.

   Example:

   event: _______ energy is changed to _______ energy

   1. television is on: electrical energy is being changed to sound (i.e., mechanical) energy.

   Assignment due: __________

2. Use the correct units of measure for force, mass, acceleration, distance, and velocity, to explain why "energy" in Albert Einstein's famous formula, $E=mc^2$, is measured in joules. *Hint:* In this famous formula, the letter "c" stands for the velocity of light.

   Assignment due: __________

_________________________ Student's Signature

_________________________ Parent's Signature

____/____/____ Date

PS13 Lesson #1

ENERGY

Work Date: ____/____/____

LESSON OBJECTIVE

Students will set up a pendulum and discuss how potential and kinetic energy can be transformed from one to the other.

Classroom Activities

On Your Mark!

Ask students to discuss the meaning of the term **energy** and define energy, at the end of their brief discussion, as the ability to do work. Ask them to imagine themselves in a car or on a roller coaster and to discuss whether or not the amount of energy they have changes as a result of their motion. Discuss and define the terms **potential** and **kinetic energy** as they appear on the student's Fact Sheet. Show students how to use the formula **G.P.E. = mgh** to determine the **gravitational potential energy** of any object in the gravitational field of a much larger object (i.e., you and the earth).

Get Set!

Show students how to set up their pendulum and lead a discussion of how **Galileo Galilei** (b. 1564; d. 1642) used a pendulum to time many of his experiments on falling objects. He did this after realizing that the mass of the "bob" suspended from the pendulum string was irrelevant to the pendulum's period of swing—one period being a full back-and-forth motion of the pendulum. Galileo discovered that it was the length of the pendulum that determined its period of swing. His discoveries led to the invention of the **pendulum clock** by **Christian Huygens** (b. 1629; d. 1695) in 1657.

Go!

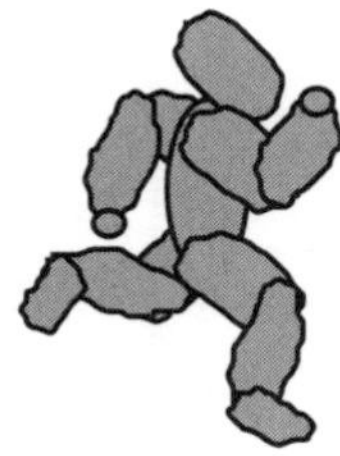

Have students perform the experiment described in Figure A on Journal Sheet #1 and conclude the lesson with a discussion of the forces influencing the pendulum's swing (i.e., potential and kinetic energy changing from one to the other as well as the force of friction impeding the motion of the pendulum). Students will discover after making careful measurements of each pendulum's period of swing that the period lengthens with the length of the pendulum. Changing the mass at the end of the pendulum does not influence the period of swing. Time permitting, you can introduce students to the two **Laws of Thermodynamics** discussed in the Teacher's Agenda and Content Notes.

Materials

ring stand and clamps, string, brass weights, a stopwatch (or rubber stopper suspended from a 15-cm length of string to act as a "pendulum timer")

Name: ______________________ Period: ________ Date: ___/___/___

PS13 Journal Sheet #1

ENERGY

FIGURE A

Directions: (1) Set up the pendulum as shown so that it can swing freely over the end of a counter or desk. (2) Use a stopwatch (or pendulum timer—i.e., a rubber stopper suspended from a 15-cm long string) to time the back-and-forth swing of different brass weights tied to the ends of different lengths of swing. (3) Round off times to the nearest full second and record the information you gather in TABLE A.

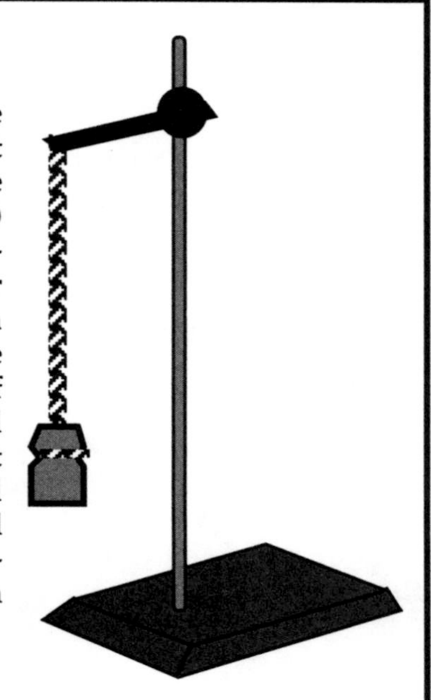

TABLE A

mass of brass weight	length of string	time to complete 10 full back-and-forth swings	average time to complete 1 full back-and-forth swing
10 grams	20 cm		
	50 cm		
	100 cm		
20 grams	20 cm		
	50 cm		
	100 cm		
50 grams	20 cm		
	50 cm		
	100 cm		

PS13 Lesson #2

ENERGY

Work Date: ___/___/___

LESSON OBJECTIVE

Students will construct a model of a steam turbine.

Classroom Activities

On Your Mark!

Prepare the rubber stopper assembly shown in Illustration A before the start of class. Ask students to name and describe the different types of engines they have seen. Tell them about the inventors, **T. Savery** and **J. Watt**, mentioned in the Teacher's Agenda and Content Notes. Explain that a steam turbine is used to generate electricity at modern electric plants. At an electric plant fuel (i.e., coal, nuclear material) is used to heat water into steam which spins a turbine that rotates a magnet. The magnet is placed inside a giant coil of wire. As it spins, the electrons inside the coil are forced to move, thus generating an electric current. Flowing wind and water can also be used, instead of steam, to rotate the magnet in an electric generator.

Get Set!

Prepare the rubber stopper assembly shown in Illustration A before the start of class and preface the construction of the steam turbine shown in Figure B of Journal Sheet #2 with the General Safety Precautions mentioned there. Draw the diagram shown in Figure B on the board and have students label all of the parts of the set-up to familiarize them with its construction. Assist students in constructing the "pinwheel" turbine to be used with this set-up.

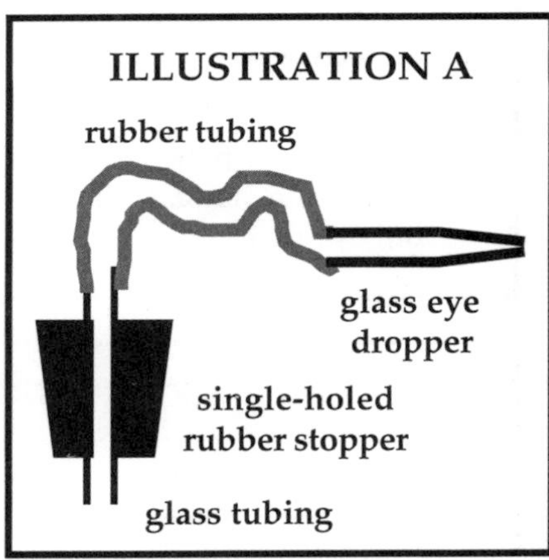

Go!

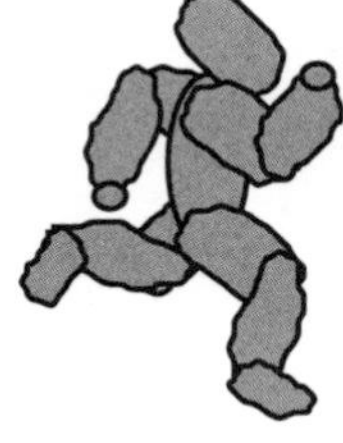

Give students ample time to construct their steam turbine. Have students list all of the energy transformations that took place in getting the chemical gas called methane (i.e., chemical potential energy) coming out of the Bunsen burner to be transformed into the motion of the pinwheel (i.e., kinetic or mechanical energy). Conclude the lesson by discussing the technical difficulties they had in building their device. Ask them if they think real engineers have similar difficulties building engines and getting them to work properly.

Materials

ring stand and clamps, Bunsen burners, rubber tubing, single-holed rubber stoppers, glass tubing, glass eye droppers, construction paper, thumb tacks, pencils

Name: ______________________ Period:________ Date: ___/___/___

PS13 Journal Sheet #2

ENERGY

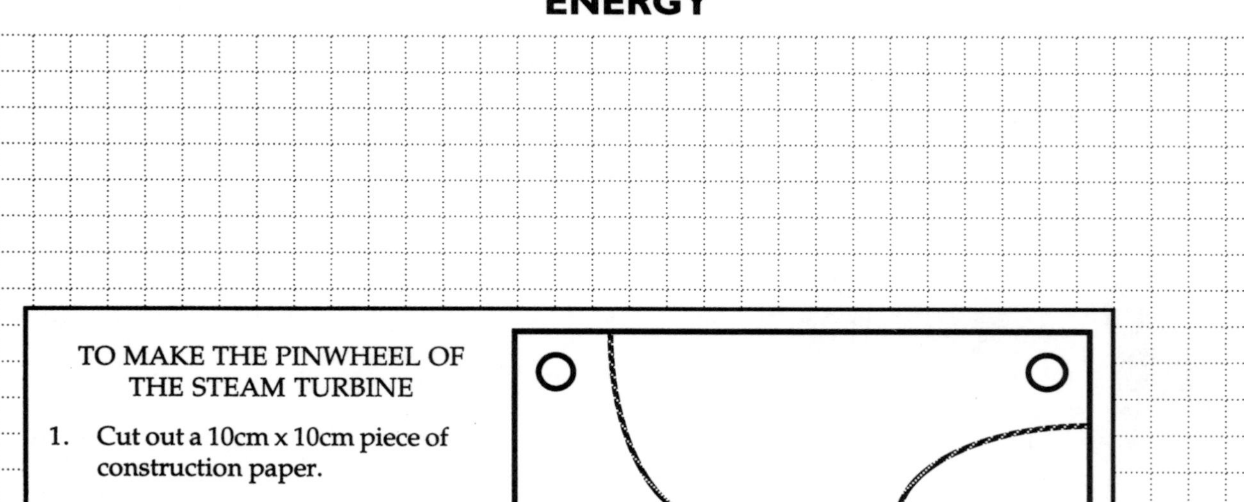

TO MAKE THE PINWHEEL OF THE STEAM TURBINE

1. Cut out a 10cm x 10cm piece of construction paper.
2. Punch holes and cut along dotted lines as shown.
3. Fold the corner holes over the center hole and press a thumb tack through the holes into a pencil. The thumb tack should be secure but still allow the pinwheel to spin freely.
4. Secure your pencil with a clamp to the ring stand at the end of the eye dropper.

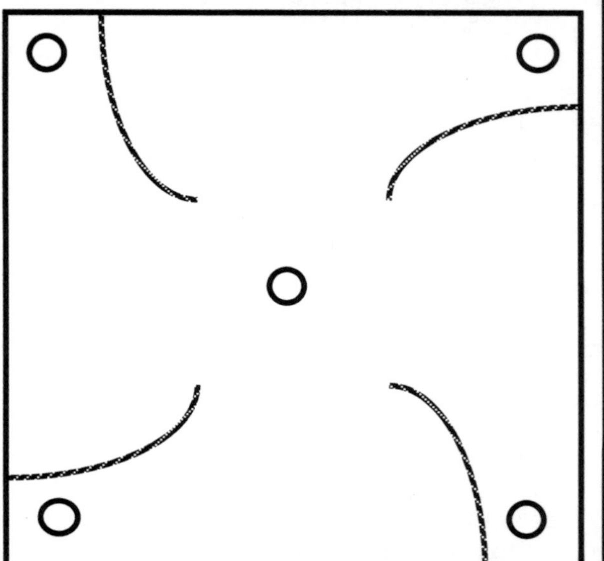

FIGURE B

Directions: (1) Pour 50 milliliters of water into an Ehrlenmeyer flask. (2) Place the flask onto a ring stand and secure it with a clamp so that it cannot be toppled. (3) Plug the flask with the rubber stopper, tubing and eye dropper assembly provided by your instructor. (4) Clamp the eye dropper and the pencil-pinwheel assembly in place as shown. (5) Turn on the Bunsen Burner and wait until steam gets the turbine going.

GENERAL SAFETY PRECAUTIONS

Be sure you are familiar with the proper use of a Bunsen burner. Wear goggles to protect your skin and eyes from being burned by SCALDING HOT STEAM. Do not touch any part of the equipment without heat resistant gloves or tongs. Clean up when the apparatus is cool.

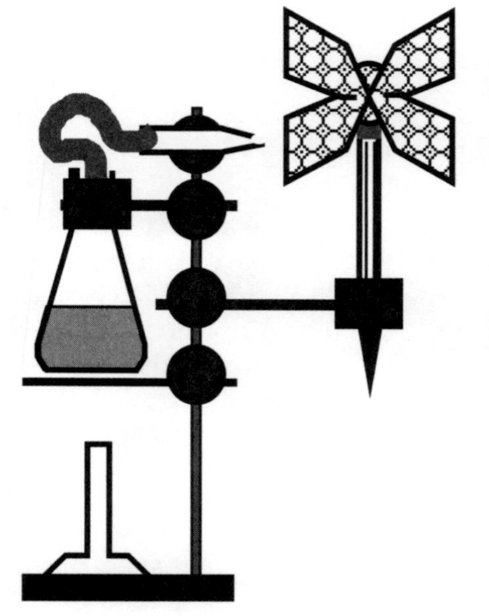

PS13 **Lesson #3**

ENERGY

Work Date: ___/___/___

LESSON OBJECTIVE

Students will explain the purpose of each stroke of the standard 4-stroke internal combustion engine used in most modern automobiles.

Classroom Activities

On Your Mark!

Give students a brief lecture of the history of the automobile. Although the American automobile manufaturer **Henry Ford** (b. 1863; d. 1947) was responsible for making the first "mass produced" auto called the **Model T**, the invention of the automobile was the result of work by many scientists and inventors. The first gas powered internal combustion engine was designed by the Belgian engineer **Jean Joseph Lenoir** (b. 1822; d. 1900) in 1860. In 1876, Lenoir's engine was improved by German engineer **Nikolaus August Otto** (b. 1832; d. 1891) whose 4-stroke engine surpassed the efficiency of Lenoir's machine. To this day, the 4-stroke cycle of the familiar internal combustion engine is called the **Otto cycle**. The first four-wheeled automobiles were designed and built by **Karl Friedrich Benz** (b. 1844; d. 1929) in the late 1880s. Automobiles built by Benz's company were sturdy, light, and cruised at the remarkable speed of 24 kilometers per hour (about 15 mph). In 1895, Frenchman **André Michelin** (b. 1853; d. 1931) began supplying automobile manufacturers in Europe with his new **pneumatic tires** which were filled with air instead of solid rubber.

Get Set!

ILLUSTRATION B

Explain that the pistons moving in and out of the combustion cylinders are connected to a crank shaft like the one shown in Illustration B. The first stroke is called the **intake stroke**. During this stroke of the downward moving piston a valve opens to admit a mixture of vaporized gasoline and air. The valve closes and the piston moves upward to compress the gas/air mixture. This second stroke is called the **compression stroke**. In the third stroke, called the **power stroke**, a **spark plug** charges the vapor with an electrical spark causing the volatile fuels inside the cylinder to explode. At the start of the fourth stroke, called the **exhaust stroke**, another valve opens and the burnt gases are flushed from the cylinder by the upward movement of the piston.

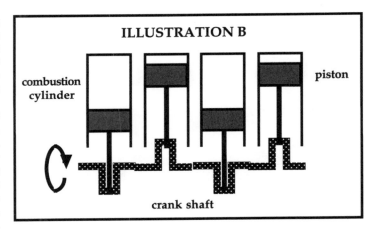

Go!

Instruct students to take careful notes during lecture and discussion and to fill out the diagram in Figure C on Journal Sheet #3. After they have finished, ask them to discuss the forces that would reduce the efficiency of this kind of engine. Ask them to consider how automobiles have affected the condition of earth's atmosphere since their invention more than a century ago.

Materials

Journal Sheet #3

Name: ______________________________ Period:________ Date: ____/____/____

PS13 Journal Sheet #3

ENERGY

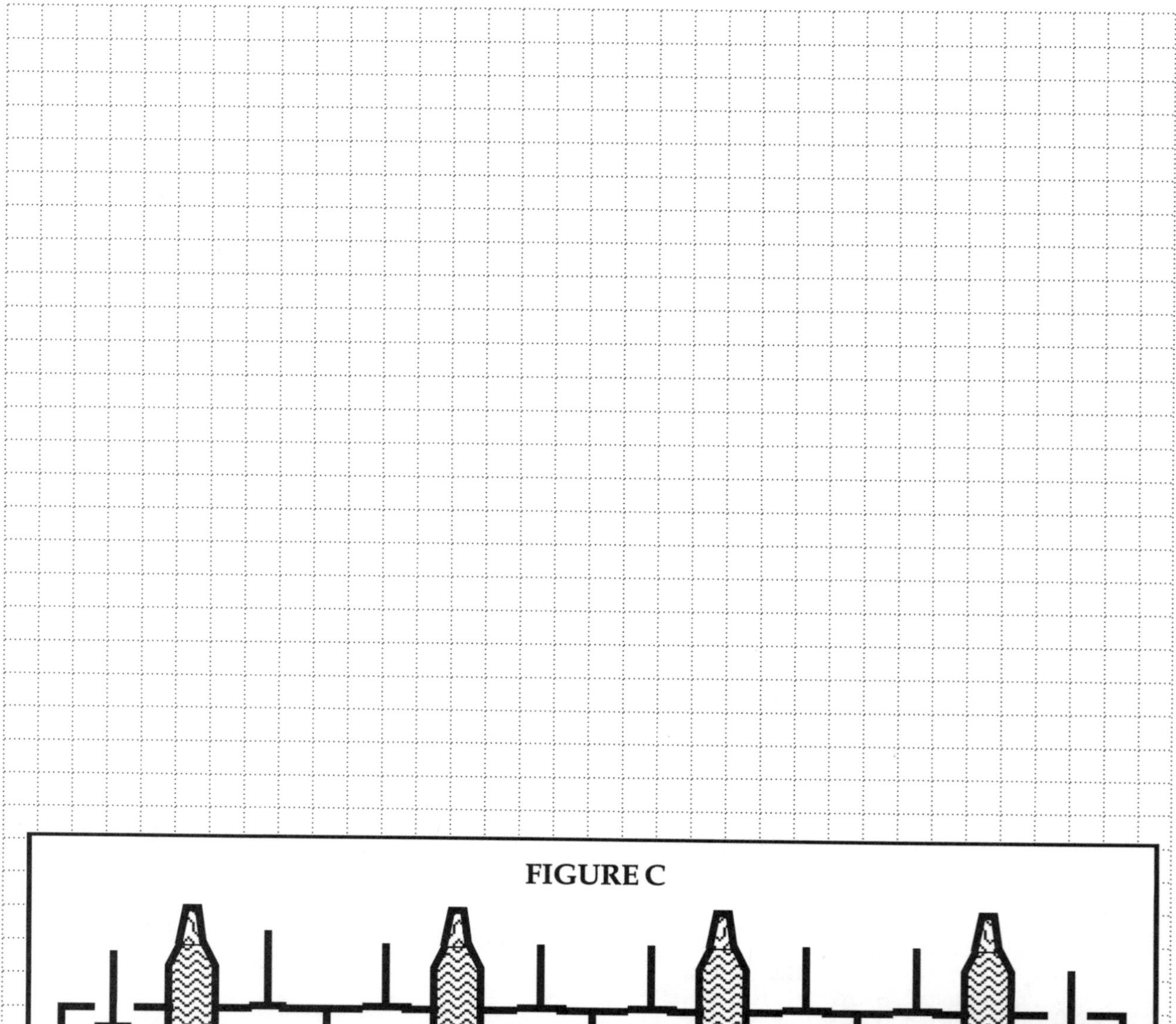

FIGURE C

PS13 **Lesson #4**

ENERGY

Work Date: ____/____/____

LESSON OBJECTIVE

Students will pick out pertinent information from an electric bill.

Classroom Activities

On Your Mark!

Remind students of the discussion you had in Lesson #2 about how electricity is created at a power plant. Remind them of the many different sources of raw fuel that can be used to "spin the magnet inside the coil that carries the electric current." Review the Second Law of Thermodynamics which makes it impossible to have a surplus of newly created energy. Explain that all of the energy we use today has been around since the beginning of time in other "stockpiled" forms. We use machines like **generators** and **batteries** to reorganize and transform energy so that it can be put to better use.

Get Set!

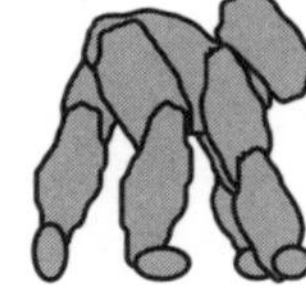

Have students list the raw sources of fuel (i.e., wood, coal, oil, nuclear material like uranium or plutonium, wind, and water whose gravitational potential energy can be stored in a dammed reservoir).

Go!

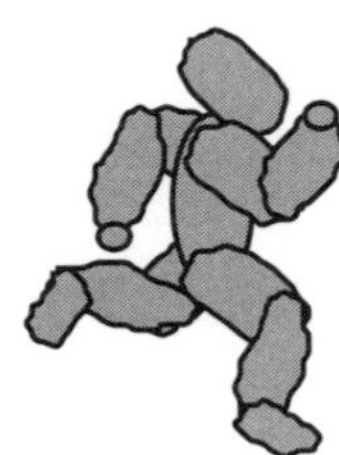

Instruct students to study the electric bill provided by you and complete the assignment on Journal Sheet #4.

Materials

old electric bill, Journal Sheet #4

Name: ______________________ **Period:**________ **Date:** ___/___/___

PS13 Journal Sheet #4

ENERGY

<u>Directions</u>: Examine the electric bill provided by your instructor and answer the questions below.

1. What is the account number of this customer? ______________________
2. What telephone number would the customer call if they had questions?
____
3. How much did the customer owe the electric company on their last bill?
____
4. What is the period covered by this bill? ______________________
5. How much money does the customer pay for each kilowatt-hour of energy used under baseline? ______________________
6. How much money does the customer pay for each kilowatt-hour of energy used over baseline? ______________________
7. How many kilowatt-hours of energy did the customer use during the period for which they are being billed? ______________________
8. Is the customer using more or less energy compared to last year? ______________________
9. Is there a state tax being paid by the customer? If there is, how much is it?
____
10. How can the customer lower his electric bill in coming months?
____
____
____

Name: ______________________ **Period:**________ **Date:** ____/____/____

PS13 Review Quiz

Directions: Keep your eyes on your own work.
Read all directions and questions carefully.
THINK BEFORE YOU ANSWER!
Watch your spelling, be neat, and do the best you can.

CLASSWORK	(~40):	_______
HOMEWORK	(~20):	_______
CURRENT EVENT	(~10):	_______
TEST	(~30):	_______
TOTAL	(~100):	_______

(A ≥ 90, B ≥ 80, C ≥ 70, D ≥ 60, F < 60)

LETTER GRADE: _______

TEACHER'S COMMENTS: ____

____

ENERGY

MULTIPLE CHOICE: Choose the letter of the word or phrase that best answers the question or completes the statement. *10 points*

_____ 1. Which of the following is a correct formula for calculating the amount of energy in a system?
(A) F • D
(B) m • c • c
(C) (m • v • v)/2
(D) m • g • h
(E) all of the above are correct formulas for energy

_____ 2. Which is the best example of potential energy?
(A) a stick of dynamite
(B) a rolling ball
(C) a surfer surfing
(D) a falling skydiver
(E) none are good examples

_____ 3. Which is the best example of kinetic energy?
(A) a pound of uranium
(B) a stalled car
(C) a boy reading
(D) an arrow in flight
(E) none are good examples

_____ 4. What happens to the potential energy of a ball rolling down a hill?
(A) it increases
(B) it decreases
(C) it is changed to kinetic energy
(D) both A and C
(E) both B and C

_____ 5. What is the correct unit of measure for energy?
(A) joule
(B) erg
(C) newton-meter
(D) foot-pound
(E) all of the above

TRUE–FALSE FILL-IN: If the statement is true, write the word TRUE. If the statement is false, change the underlined word to make the statement true. *10 points*

________________ 6. Energy is the ability to have power.

________________ 7. Energy of motion is called kinetic energy.

________________ 8. Energy of composition or position is called potential energy.

________________ 9. In his famous formula, $E=mc^2$, Sir Isaac Newton explained that small amounts of matter can release large amounts of energy.

________________ 10. Energy cannot be transformed from one form to another.

PROBLEMS

Directions: Show all formulas and calculations. Be sure to use the correct unit of measure with your FINAL ANSWER. *10 points*

What is the potential energy of a 5-kilogram object that is 4 meters above the ground? *Hint*: The acceleration of gravity is 9.8 meters/second/second.

Final Answer: ________________

What is the kinetic energy of a 10,000 kilogram jet moving at 2,000 meters per second? *Hint:* Use scientific notation to make the math easier.

Final Answer: ________________

________________________ Student's Signature

________________________ Parent's Signature

____/____/____ Date

PS14

ELECTROMAGNETISM

Teacher's Classwork Agenda and Content Notes

Classwork Agenda for the Week

1. Students will trace the pattern created by a magnetic field and discuss the basic laws of magnetism.
2. Students will determine the types of materials influenced by a magnetic field.
3. Students will demonstrate that magnetism and electricity are two aspects of a single force.
4. Students will build an electromagnet.

Content Notes for Lecture and Discussion

Of the four natural forces affecting the behavior of matter (i.e., the **strong force**, the **weak force**, **gravity**, and **electromagnetism**), it is the scientist's understanding of **electromagnetism** that has most enhanced our quality of life through the development of new and useful technology.

Ancient people were aware of the attractive powers of lodestone and the curious effects of rubbing amber against pieces of wool. But it was not until the year 1600 that English scientist **William Gilbert** (b. 1540; d. 1603) elaborated on the distinctive qualities of these phenomena in his book *De magnete*. Gilbert treated the two phenomena as autonomous in nature. He saw magnetism as a "dipolar" phenomenon and attributed the startling electrical effects of rubbing wool and amber together to be the result of frictional forces. For two centuries, scientists, mathematicians, and philosophers argued the underlying causes of these two simple wonders. It was not until the invention of the **voltaic pile** in 1800 by **Count Alessandro Volta** (b. 1745; d. 1827), and the development of new technology with which to study the effects of electricity, that scientists began to question Gilbert's explanation of these phenomena.

Hans Christian Oersted's (b. 1777; d. 1851) discovery in 1820 that an electric current produced a magnetic field was the beginning of an era of theoretical breakthrough and technological advance. The forces of electricity and magnetism had been unified! It was English chemist **Michael Faraday** (b. 1791; d. 1867) who elaborated on the notion of an **electromagnetic field** which guided him to the invention of the transformer. A **transformer** is a device used to increase or decrease the strength of an electric current. Faraday's ideas were put into mathematical form by **William Thomson Kelvin** (b. 1824; d. 1907) and **James Clerk Maxwell** (b. 1831; d. 1879) whose equations suggested that **light energy** was the result of an electromagnetic process. This notion was confirmed by the work of German physicist **Heinrich Hertz** (b. 1897; d. 1854) who devised a method for detecting **electromagnetic waves**. Hertz noted that the behavior of electromagnetic waves resembled the behavior of radiation produced by luminous sources of heat and light. The technology developed as the result of this work allowed physicists **Albert Abraham Michelson** (b. 1852; d. 1931) and **Edward William Morley** (b. 1838; d. 1923) to accurately measure the speed of light and dispel the notion that electromagnetic waves were propagated through an interstellar aether. The **Michelson-Morley experiment** of 1887 allowed **Albert Einstein** (b. 1879; d. 1955) to develop his **Theory of Relativity** in 1905.

In Lesson #1, students will make the invisible visible by tracing patterns produced by iron filings around a magnet. In Lesson #2, they will examine the types of materials affected by a magnetic field. In Lesson #3, students will repeat Oersted's famous experiment and demonstrate that electricity creates a magnetic field. In Lesson #4, students will build an electromagnet. This last lesson will reinforce the notion that ideas which seem purely theoretical today might prove technologically valuable tomorrow.

PS14 Content Notes *(cont'd)*

ANSWERS TO THE HOMEWORK PROBLEMS

Answers will vary but should express the following notions:

1. The closer together two magnets facing north to north are placed, the more strongly they will repel one another.
2. The farther apart two magnets facing south to south are placed, the more weakly they will repel one another.
3. The closer together two magnets facing north to south are placed, the more strongly they will attract one another.
4. The farther apart two magnets facing north to south are placed, the more weakly they will attract one another.

ANSWERS TO THE END-OF-THE-WEEK REVIEW QUIZ

1. lodestone
2. alloys
3. electrons
4. true
5. pole
6. alike
7. unlike
8. iron and nickel
9. field
10. electromagnet
11. increases
12. compass
13. core
14. thousands
15. cannot

Students' descriptions of how to build an electromagnet will vary but should reflect the student's grasp of the activity performed in Lesson #4

Name: ________________________ **Period:**________ **Date:** ___/___/___

PS14 Fact Sheet

ELECTROMAGNETISM

CLASSWORK AGENDA FOR THE WEEK

(1) Draw the invisible magnetic field surrounding a magnet and discuss the laws of magnetism.
(2) Determine the types of materials influenced by a magnetic field.
(3) Show that magnetism and electricity are two aspects of a single force.
(4) Build an electromagnet.

There is an old story about an ancient explorer whose walking staff became stuck to the ground. Upon digging up the metal tip of his staff, the explorer discovered that it was held to the ground by a strange black rock. The rock was a form of iron able to attract objects made of metal. Since ancient times, mankind has known about the **magnetic properties** of rocks called **lodestone**. Lodestone is made of a mineral called **ferrous oxide**: an ore containing **iron** and **oxygen** atoms. Today, we can make **magnets** out of metal **alloys**. These alloys are usually mixtures of the elements *al*uminum, *ni*ckel, and *co*balt (i.e., *alnico*). The "field of **attraction** and **repulsion**" surrounding a magnet can be "seen" by surrounding the magnet with tiny bits of metal (i.e., iron filings).

Every magnet has a **north** (N) and **south** (S) **pole**. Magnetic poles—as far as we know—do not exist in isolation. A north pole, for example, cannot exist without a south pole and vice versa. Poles that are alike (i.e., N and N, S and S) **repel** one another. Poles that are not alike (i.e., N and S) **attract** one another. You may also have discovered when playing with magnets that the force pulling **unlike poles** together, or pushing **like poles** apart, increases as the magnets get closer together. The **force of magnetism** is similar to the **force of gravity** in this respect. Objects placed close together have a greater gravitational attraction for one another than when they are placed far apart. Unlike magnetism, however, gravity can only pull things together not push them apart.

Magnets are found in many different places. The **core of the earth**, which is made of iron, behaves like a magnet. As the world spins on its **axis** it produces a gigantic **magnetic field** that surrounds our planet. A **compass** needle made of magnetized metal (like alnico) points toward the North and South Poles of our planet. The needle lines up with the magnetic field surrounding the earth.

In 1820, Danish physicist **Hans Christian Oersted** (b. 1777; d. 1851) discovered quite by accident that magnetism and electricity are related. He was showing his students how to produce an electric current when a compass needle near the wires hooked to his battery changed direction. After several careful experiments Oersted concluded that the electricity passing through the wire was creating a magnetic field.

Three years later, English physicist **William Sturgeon** (b. 1783; d. 1850) wrapped an iron bar with a wire coil and made the first electromagnet. An **electromagnet** is constructed by passing electricity through a coiled piece of wire. Since moving electrons produce a magnetic field, as Oersted discovered, an iron core placed inside the coil of wire will become "magnetized." Sturgeon's electromagnet was able to lift objects 20 times its own weight. Perhaps you have seen a modern electromagnet in a junk yard lift scraps of metal weighing several thousand pounds!

Homework Directions

1. Write one or two sentences describing the amount and direction of forces between two magnets placed north-to-north as you move them closer together.

2. Write one or two sentences describing the amount and direction of forces between two magnets placed south-to-south as you move them farther apart.

3. Write one or two sentences describing the amount and direction of forces between two magnets placed north-to-south as you move them closer together.

4. Write one or two sentences describing the amount and direction of forces between two magnets placed north-to-south as you move them farther apart.

Assignment due: __________

______________________ Student's Signature

______________________ Parent's Signature

___/___/___ Date

PS14 **Lesson #1**

ELECTROMAGNETISM

Work Date: ____/____/____

LESSON OBJECTIVE

Students will trace the pattern created by a magnetic field and discuss the basic laws of magnetism.

Classroom Activities

On Your Mark!

Ask students if they have ever used a compass. Demonstrate how a compass is used and allow students to share ideas about how a compass works. Explain that the ancient Greeks used **lodestone** compasses to assist them in navigating their way around the Aegean and Adriatic Seas as early as 500 B.C. In the eleventh century, the Chinese magnetized metal needles by rubbing them against pieces of lodestone; then, they inserted the needles into pieces of straw and floated them in small bowls of water.

Get Set!

Have students copy your board drawing of Illustration A. Explain that the iron core of our planet generates a magnetic field around our world as it rotates on its axis. Tell them that solar radiation coming into our planet's atmosphere follows the magnetic lines of force at the North and South Poles and lights up the sky at higher latitudes creating the colorful *aurora borealis* or "northern lights."

ILLUSTRATION A

The iron-nickel core of the earth rotates on its axis and creates a magnetic field.

Go!

Have students perform the experiment described in Journal Sheet #1. Conclude the activity by having students copy the two laws of magnetism: (1) Magnetic poles do not exist in isolation. All magnets have a north and south pole. (2) Like magnetic poles repel. Unlike magnetic poles attract.

Materials

compasses, bar magnets, horseshoe magnets, Journal Sheet #1

Name: ______________________________ Period:________ Date: ___/___/___

PS14 Journal Sheet #1

ELECTROMAGNETISM

Directions: (1) Trace the magnets given to you by your instructor in the following three arrangements: (a) north pole facing north pole, (b) south pole facing south pole, and (c) north pole facing south pole. Be sure to label the north and south poles of the magnets in each drawing. (2) Place the magnets underneath this Journal Sheet so that they are directly beneath the drawings. (3) Sprinkle a teaspoon of iron filings on the Journal Sheet and gently spread the filings over each drawing. Then, (4) trace the pattern made by the filings with a pencil.

Describe the pattern of the iron filings around each arrangement of magnets:

(a) north to north: ______________________________

(b) south to south: ______________________________

(c) north to south: ______________________________

What can you conclude about the way magnets behave given the arrangement of the magnetic field around the magnets in each situation?______________________________

____

____

____

PS14 **Lesson #2**

ELECTROMAGNETISM

Work Date: ____/____/____

LESSON OBJECTIVE

Students will determine the types of materials influenced by a magnetic field.

Classroom Activities

On Your Mark!

Construct the set-up shown in Figure A on Journal Sheet #2. Ask students to discuss the forces holding up the paper clip. Ask them if the forces at work can be altered in any way.

Get Set!

Give students time to construct the same set-up at their tables or desks.

Go!

Have students perform the activity described on Journal Sheet #2. When they are finished, challenge them to hang a second paper clip from the first to see if the first paper clip has become magnetized. If the bar magnet is strong enough this will be possible. Explain the concept of **electromagnetic induction**. Electromagnetic induction is the production of a magnetic or electric effect by one substance in another without the two bodies touching.

Materials

ring stand and clamps, paper clips, brass or other weights, fine thread, bar or horseshoe magnets

Name: ______________________ Period:________ Date: ____/____/____

PS14 Journal Sheet #2

ELECTROMAGNETISM

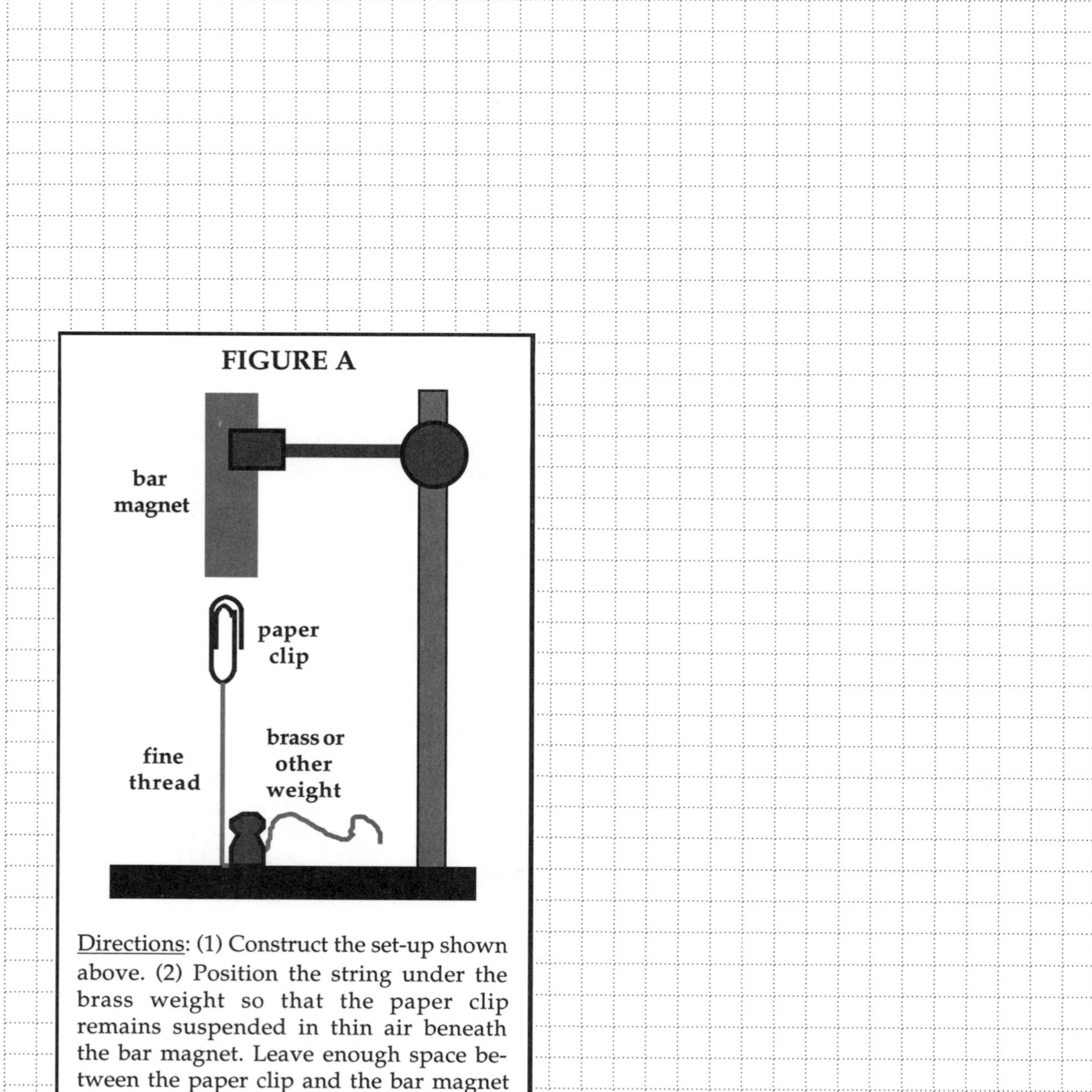

<u>Directions</u>: (1) Construct the set-up shown above. (2) Position the string under the brass weight so that the paper clip remains suspended in thin air beneath the bar magnet. Leave enough space between the paper clip and the bar magnet to pass a variety of materials (i.e., a wooden ruler, a plastic ruler, a glass rod, paper, pens, needles and pins). (3) Make a chart to record whether or not the suspended paper clip moved as you passed each item between it and the bar magnet. *Be careful not to touch the paper clip directly.*

PS14 Lesson #3

ELECTROMAGNETISM

Work Date: ____/____/____

LESSON OBJECTIVE

Students will demonstrate that magnetism and electricity are two aspects of a single force.

Classroom Activities

On Your Mark!

List the **four forces of nature** on the board and describe the character of each. Have students copy this information onto Journal Sheet #3. The **strong force** is the force that holds the nuclei of atoms together. As students may have learned (or will learn) in chemistry, atoms have a **nucleus** made of **subatomic particles** called **protons** and **neutrons**. Protons have a **positive (+) electric charge**. Like the "like poles of a magnet," positive electric charges repel one another. At extremely small distances of approximately 10^{-13} meters (one ten-trillionth of a meter) the strong force overcomes the mutual repulsion between protons and glues the nucleus together. The **weak force** is the force that sometimes "fools" the strong force, allowing radioactive particles to escape from the nucleus. **Gravity**, of course, is the force of attraction between all forms of matter. **Electromagnetism** is the force of attraction or repulsion produced by the vibration of **electrons**. Electrons are the tiny subatomic particles swirling in a cloud around the nuclei of atoms.

Get Set!

Use the information provided in the Teacher's Agenda and Content Notes to give students the historical events that led to **Hans Christian Oersted**'s (b. 1777; d. 1851) discovery of the relationship between magnetism and electricity.

Go!

After reviewing the General Safety Precautions outlined in Figure B of Journal Sheet #3, assist students in constructing the electric circuit shown in Figure B and performing the experiment described there. Conclude the lesson by showing students the **Left Hand Rule** that describes the orientation of an electromagnetic field shown in Illustration B.

ILLUSTRATION

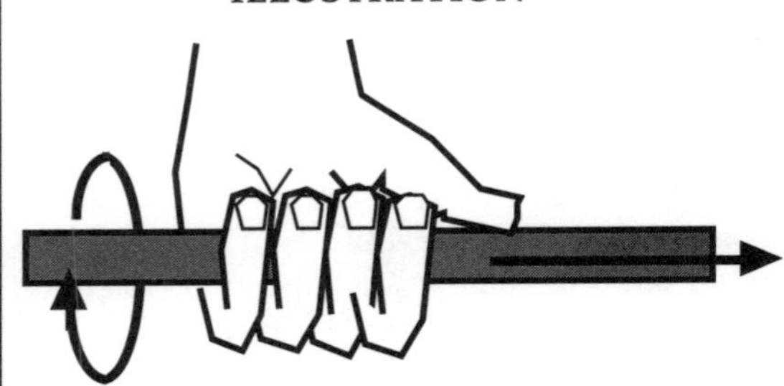

The Left Hand Rule

Wrap your left hand around a wire with the thumb pointing in the direction that the current is flowing. Electric current flows from the negative (-) side of the battery through the electric circuit to the positive (+) side of the battery. The curl of the fingers around the wire shows the orientation of the magnetic field surrounding the wire with respect to the electric field flowing through the wire. Note: The magnetic field is perpendicular to the electric field.

Materials

D-cell batteries, battery holders and alligator clips (or tape), insulated thin-guage copper wire, switches, compasses

Name: ______________________ Period:________ Date: ____/____/____

PS14 JOURNAL SHEET #3

ELECTROMAGNETISM

FIGURE B

Directions: (1) Construct the electric circuit shown below using the equipment given to you by your instructor. (2) Position the wire over the compass, pointing the wire due north as indicated by the compass. Your wire should be directly above the compass running parallel to the needle. (3) Observe what happens when you close the switch. (4) Record what you observe as you move the compass under the wire from the battery to the switch.

GENERAL SAFETY PRECAUTIONS

Open the switch if the wire begins to get hot and as soon as you are finished making your observations.

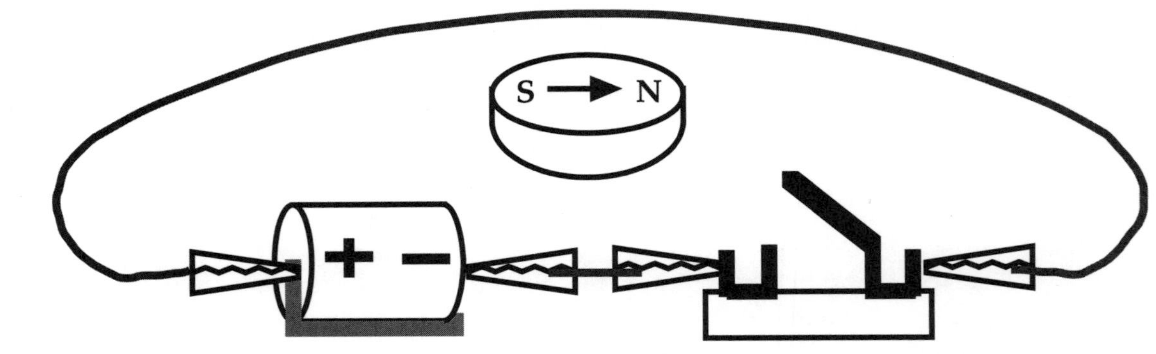

PS14 **Lesson #4**

ELECTROMAGNETISM

Work Date: ____/____/____

LESSON OBJECTIVE

Students will build an electromagnet.

Classroom Activities

On Your Mark!

Review the Left Hand Rule discovered in Lesson #3.

Get Set!

Discuss the events leading up to the invention of the **electromagnet** by **William Sturgeon** (b. 1783; d. 1850) in 1823. Inform students that the key to constructing a strong electromagnet is to make sure that the wire coils are neatly wrapped around the iron core. The coils must be tight and straight and must not overlap. The more coils that students can fit neatly around the two taped nails the stonger their magnet will be.

Go!

After reviewing the General Safety Precautions outlined in Figure C of Journal Sheet #4, assist students in constructing the electromagnet shown in Illustration C on Journal Sheet #4.

Materials

D-cell batteries, battery holders and alligator clips (or tape), insulated thin-gauge copper wire, switches, compasses, iron nails, tape

Name: ______________________ **Period:** ________ **Date:** ___/___/___

PS14 Journal Sheet #4

ELECTROMAGNETISM

FIGURE C

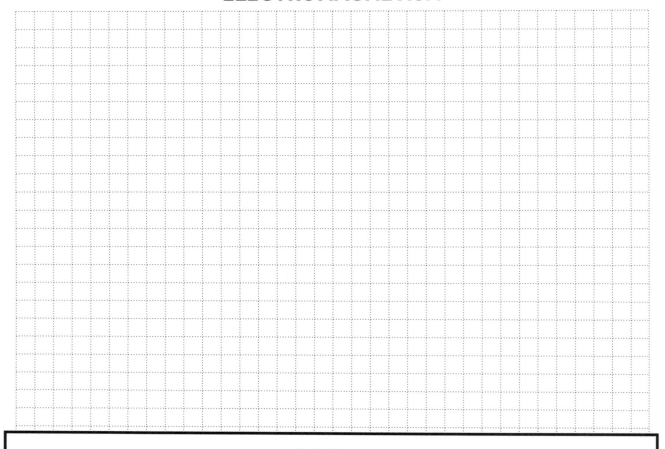

Directions: (1) Construct the electromagnet shown below with the equipment given to you by your instructor. Follow your instructor's directions precisely. (2) Test your electromagnet to see if it can pick up a single paper clip. (3) Is it strong enough to pick up more than one paper clip? (3) Record what happens when you touch one paper clip to another already touching the electromagnet.

GENERAL SAFETY PRECAUTIONS

Open the switch if the wire begins to get hot and as soon as you are finished making your observations.

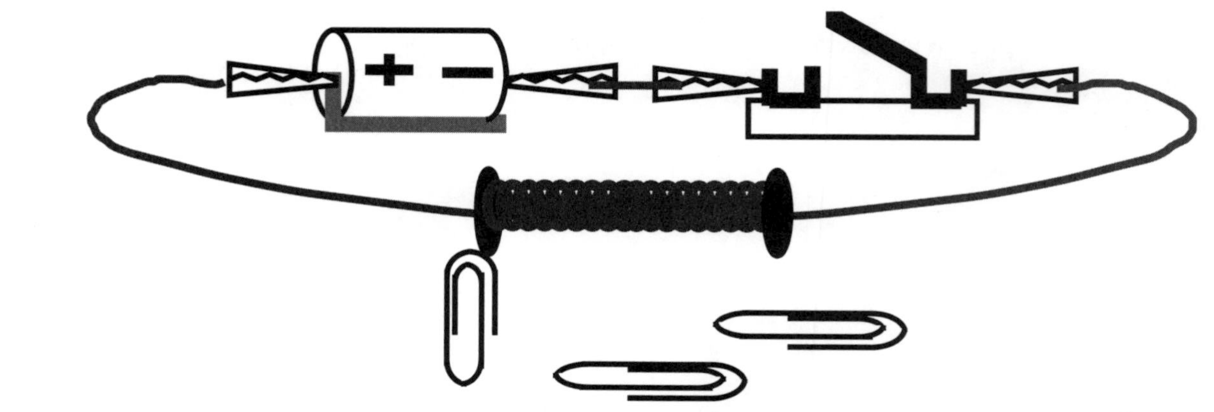

Name: ______________________________ **Period:**________ **Date:** ____/____/____

PS14 Review Quiz

Directions: Keep your eyes on your own work.
Read all directions and questions carefully.
THINK BEFORE YOU ANSWER!
Watch your spelling, be neat, and do the best you can.

CLASSWORK	(~40):	_______
HOMEWORK	(~20):	_______
CURRENT EVENT	(~10):	_______
TEST	(~30):	_______
TOTAL	(~100):	_______

(A ≥ 90, B ≥ 80, C ≥ 70, D ≥ 60, F < 60)

LETTER GRADE: _______

TEACHER'S COMMENTS: ____

____

ELECTROMAGNETISM

TRUE–FALSE FILL-IN: If the statement is true, write the word TRUE. If the statement is false, change the underlined word to make the statement true. *20 points*

_______________ 1. Natural magnets are made of rocks called alloys.

_______________ 2. Manmade magnets can be made from metal lodestone.

_______________ 3. Magnetism is caused by the "lining up" of protons that spin in the same direction.

_______________ 4. Moving electrons in a metal wire can create magnetism.

_______________ 5. Every magnet has a north (N) and south (S) side.

_______________ 6. Magnetic poles that are not alike repel one another.

_______________ 7. Magnetic poles that are alike attract one another.

_______________ 8. The core of the earth is made of alnico.

_______________ 9. As the world spins on its axis it produces a gigantic wind that surrounds our planet.

_______________ 10. A(n) generator uses electricity traveling through a coiled piece of wire to magnetize a piece of iron surrounded by the wire.

SENTENCE COMPLETION: Fill in the blank with the word or phrase that best completes the sentence. *5 points*

11. Moving two magnets closer to one another _______________ the amount of force between the magnets.
12. A(n) _______________ uses a magnet to indicate direction.
13. The earth's _______________ behaves like a giant magnet.
14. Mankind has used magnets for _______________ of years.
15. A magnetic field _______________ be seen directly.

ESSAY

Directions: Write a short procedure for constructing an electromagnet. *5 points*

Materials Needed: ______________________________

____

____

How to build an electromagnet: ______________________________

____

____

____

____

____

____

______________________________ Student's Signature

______________________________ Parent's Signature

____/____/____ Date

PS15

STATIC AND CURRENT ELECTRICITY

TEACHER'S CLASSWORK AGENDA AND CONTENT NOTES

Classwork Agenda for the Week

1. Students will create a static electric field that can be detected with the use of an electroscope.
2. Students will draw electrical circuit diagrams for a simple series circuit and explain the relationship between amperage, voltage, and resistance in a circuit.
3. Students will construct a simple parallel circuit and discuss the advantages of this kind of circuit over a series circuit.
4. Students will construct a simple electric motor.

Content Notes for Lecture and Discussion

The technological revolution of the nineteenth and twentieth centuries has its roots in the theoretical and practical research conducted by scientists in the eighteenth century. **Benjamin Franklin** (b. 1706; d. 1790) demonstrated the existence of natural **static electricity** in 1752 in his now famous "key on a kite tail experiment." His research led to the invention of the **lightning rod**; and ever since, lightning rods have been used to protect manmade structures from the devastating effects of lightning by discharging the electricity created during violent thunderstorms to the ground. In 1800, **Count Alessandro Volta** (b. 1745; d. 1827) designed and built the first **wet cell battery** to produce an **electric current**. He alternately stacked metal plates of silver and zinc, separating them with cloth pads soaked in salt water or acid. This is the essential design of any battery. Metals immersed in an electrolyte will separate and stockpile electric charges. An **electrolyte** is any substance that causes the dissociation of atoms into **positive** and **negative ions**. The electrolyte may be a liquid, as it is it in a car battery, or a paste, as it is in any common **dry cell battery**. The first dry cell battery was made by the French engineer **Georges Leclanché** (b. 1839; d. 1882). Leclanché's battery used carbon and zinc rods immersed in an electrolytic paste composed of starch, carbon powder, ammonium chloride and manganese dioxide. The concurrent and later discoveries of scientists **Coulomb**, **Ampere**, **Ohm**, **Oersted**, **Sturgeon**, and **Faraday**—to name a few—made the science of electronics the foundation of modern technology.

In Lesson #1 students will create a **static electric field** by constructing an **electroscope**. The aluminum foil "flags" inside the electroscope become depositories of electric charges created by touching charged objects (i.e., a balloon that has been rubbed against one's sweater) to the electroscope rod. If your science department has a **van de Graaff generator** you can use it to impress students with the amount of static electricity that can be created by the force of friction. Some van de Graaff generators can generate more than several million volts. The van de Graaff generator was invented by American physicist **Robert Jemison van de Graaff** (b. 1901; d. 1967) in 1929. Since then, the device has been used to create and accelerate charged particles for experimental study.

In Lessons #2 and #3, students will learn how to use **Ohm's Law** in calculating the amount of **amperage (I)**, **voltage (V)**, and **resistance (R)** in an electric circuit. Ohm's Law expresses the relationship between amperage, voltage, and resistance in the following formula: $\mathbf{I = V \div R}$. The equation expresses the notion that there is a direct relationship between amperage and voltage and an inverse relationship between amperage and resistance. Students will construct and compare a number of **series** and **parallel circuits**. They will discover that parallel circuits have several advantages over series circuits and be able to explain why the appliances in their homes are hooked up in parallel and not series.

In Lesson #4 students will construct a simple electric motor using an electromagnet similar to the one they built in the unit entitled "Electromagnetism." The construction of this motor requires some manual dexterity and will impress students with the important relationship between theoretical and practical science. They will realize that skill, attention to craft, and the use of appropriate materials are as important as the understanding of concepts and ideas when it comes to creating new and innovative technology.

ANSWERS TO THE HOMEWORK PROBLEMS

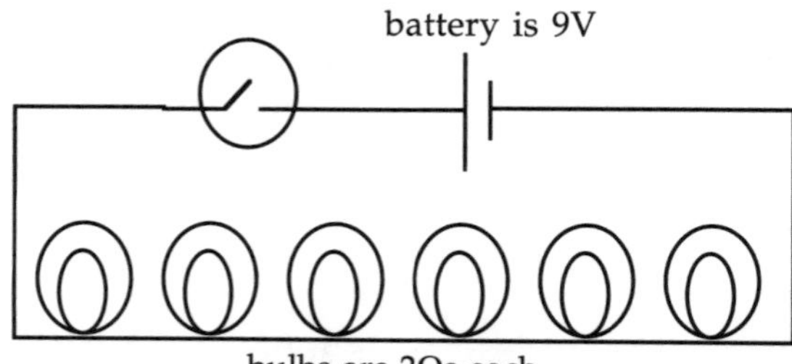

bulbs are 2Ωs each

PROBLEM #1

Ohm's Law applied to a series circuit where the total resistance (R_t)is the sum of the individual resistances:

$$I = V \div R$$
$$= 9V \div (6 \text{ bulbs} \times 2\Omega)$$
$$= 9V \div 12\Omega$$
$$= \underline{0.75 \text{ amps}}$$

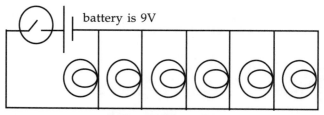

battery is 9V

bulbs are 2Ωs each

PROBLEM #2

Ohm's Law applied to a parallel circuit where the total resistance (R_t) is the "inverse" of sum of the "inverses" of the individual resistances:

$$\frac{1}{R_t} = \frac{1}{R_1} + \frac{1}{R_2} + \frac{1}{R_3} \ldots \text{etc.}$$

Therefore, the total resistance of the bulbs in this parallel circuit is:

$$\frac{1}{R_t} = \frac{1}{2\Omega} + \frac{1}{2\Omega} + \frac{1}{2\Omega} + \frac{1}{2\Omega} + \frac{1}{2\Omega} + \frac{1}{2\Omega} = \frac{6}{2\Omega}$$

So, $R_t = 2\Omega \div 6 = \frac{1}{3}\Omega$

$$I = V \div R$$
$$= 9V \div \frac{1}{3}\Omega = \underline{27 \text{ amps}}$$

ANSWERS TO THE END-OF-THE-WEEK REVIEW QUIZ

1. Volta
2. coulombs
3. amperes
4. voltage
5. true
6. Ohm's
7. true
8. iron and nickel
9. motor
10. generator
11. series
12. They would go out.
13. 6Ω
14. brighter
15. I = V ÷ R = 12V ÷ 6Ω = 2 amps

Students' list of scientists and inventors will vary but should reflect their grasp of Fact Sheet and lecture notes.

Name: ______________________________ Period:________ Date: ____/____/____

PS15 Fact Sheet

STATIC AND CURRENT ELECTRICITY

CLASSWORK AGENDA FOR THE WEEK

(1) Create a "static" electric field.
(2) Construct series circuits.
(3) Construct parallel circuits.
(4) Build an electric motor.

Count Alesandro Volta (b. 1745; d. 1827) invented the first battery in 1800. His invention became known as the **voltaic pile**. A **battery** is a chemical tool for separating stores of **positive (+)** and **negative (-) electric charges**. The electric charges in a battery are the product of the **subatomic particles** inside the **atoms** of the chemical molecules that go into the battery. Atoms have a center called a **nucleus** which contains positively charged particles called **protons**. Swirling around the nucleus of an atom are the atom's **electrons** which have a negative charge.

Electric charge is measured in **coulombs** after the French scientist **Charles Augustin de Coulomb** (b. 1736; d. 1806). Coulomb was the first to measure the amount of force between electric charges. Electric charges can flow from one storage site to another by connecting the sites with a metal wire usually made of copper. The flow of electric charges is called a **current**. Current is measured in **amperes** after the French physicist **André Marie Ampere** (b. 1775; d. 1836). Ampere is famous for discovering the **Left Hand Rule** that describes the direction of a magnetic field around a wire carrying an electric current. The amount of force required to "push" electric charges through a wire is called **voltage**. Voltage is measured in **volts**. As electric charges pass through a wire, the wire heats up due to friction between charged particles. The heat interferes with the flow of electric current. The wire's **resistance** to the flow of electricity is measured in **ohms** after the German physicist **Georg Simon Ohm** (b. 1789; d. 1854).

The relationship between amperage, voltage, and resistance can be expressed using the following formula called **Ohm's Law**:

$$\mathbf{I = V \div R}$$

where **I** is amperage, **V** is voltage, and **R** is resistance.

There are a variety of ways to hook up the wires of an electric circuit in order to get electricity to flow through a circuit connected to appliances. A circuit requires two simple things. First, it needs a pathway of **conductors** made of material through which electrons can move. Second, a circuit needs **insulators** made of materials that prevent the electrons from going astray. Today, most circuits do not have wires at all. They are fused onto **solid state circuit boards** or **silicon chips**. Silicon chips are made of carefully positioned atoms arranged in crystals. The first silicon chips were made by an American company named **Texas Instruments** in 1958. These tiny silicon crystals—made of much the same substance as plain sand—are the foundation of the modern computer industry.

A **motor** is an appliance that is built using an electromagnet. The first motor was invented in 1821 by English chemist **Michael Faraday** (b. 1791; d. 1867). A motor works by reversing the positions of the electromagnet's north and south poles. Putting one pole of a fixed magnet next to the "changing" poles of the electromagnet causes the electromagnet to spin, as "like" poles repel one another: north from north then south from south. A motor transforms electrical energy to kinetic energy.

A **generator** is the opposite of a motor. A generator transforms kinetic energy to electrical energy. The first practical generator was used by **Thomas Alva Edison** (b. 1847; d. 1931) in the 1880s. At a power plant that produces electricity for use by many people, fuels like coal, natural gas, petroleum, or uranium are used to heat and boil water to make steam. Under pressure, the steam is forced against a wheel called a **turbine**. The turbine spins a magnet inside a coil of metal wire. The spinning magnet causes electrons to flow in the wire to produce electricity. Wind and water can also be used to spin the magnet inside the wire coil.

Homework Directions

1. Using the symbols introduced in class, draw a series circuit showing a 9-volt battery lighting six 2-ohm light bulbs. Label your drawing and tell how much amperage is flowing through the circuit.

2. Using the symbols introduced in class, draw a parallel circuit showing a 9-volt battery lighting six 2-ohm light bulbs. Label your drawing and tell how much amperage is flowing through the circuit.

Assignment due: __________

_________________________ Student's Signature

_________________________ Parent's Signature

____/____/____ Date

STATIC AND CURRENT ELECTRICITY

Work Date: ____/____/____

LESSON OBJECTIVE

Students will create a static electric field that can be detected with the use of an electroscope.

Classroom Activities

On Your Mark!

If your science department has a **van de Graaff generator** use it to demonstrate how static electric charges can be created. Exercising appropriate safety precautions, use the generator to demonstrate the standard classroom phenomena (i.e., the hair on a person touching the generator will stand on end). If a van de Graaff is not available, you can perform the following experiment: (1) Inflate two balloons and suspend them from the ceiling with string, about one foot apart from each other. (2) Rub one of the balloons with a piece of wool from an old sweater. (3) Have students note the fact that the balloons are attracted to one another. (4) Rub both balloons with the same piece of wool. (5) Have students note the fact that the two balloons now repel each other. *Conclusion:* When you treat the balloons differently (i.e., only one was rubbed with the wool), there was an "attractive force" between them. When you treated both of them equally (i.e., rubbed both balloons with the wool), there was a "force of repulsion" between them. Introduce them to the following **rule of electrostatics**: "Unlike electric charges attract, while like electric charges repel." Remind them of the first Law of Magnetism which states essentially the same thing about magnetic poles.

Get Set!

Discuss the accomplishments of the scientists mentioned in the Fact Sheet and the Teacher's Agenda and Content Notes. Specifically the experiments of **Benjamin Franklin** (b. 1706; d. 1790) and the discovery of **Charles Augustin de Coulomb** (b. 1736; d. 1806) who measured the amount of force between static electric charges. Have them copy the following definition on Journal Sheet #1:

coulomb: unit of electrical charge equal to the charge on 6×10^{23} electrons.

Go!

Prepare the nail-in-rubber stopper assembly before the start of class. Show students how to construct their **electroscope** and complete the activity by following the directions in Figure A on Journal Sheet #1. They may also observe the effects of their statically charged balloon when placed near the hair on their arms or tiny bits of paper and alluminum foil.

Materials

Ehrlenmeyer flasks, single-holed rubber stoppers, iron nails, "twistie-ties" used to wrap bread/garbage bags, thin sheets of aluminum foil

Name: ______________________________ **Period:**________ **Date:** ___/___/___

PS15 Journal Sheet #1

STATIC AND CURRENT ELECTRICITY

FIGURE A

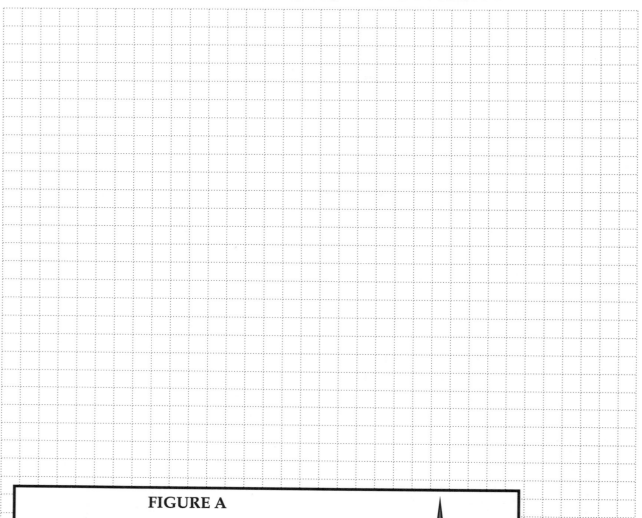

Directions: (1) Strip the paper/plastic covering off a metal "twistie-tie" normally used to tie up items like bread and garbage bags. (2) Wrap the wire around the head of the nail stuck in the rubber stopper given to you by your instructor. (3) Cut small rectangular strips of thin aluminum foil and rub the foil flat. (4) Punch a small hole in the end of each piece of foil and hang them loosely from the ends of the wire. (5) Carefully insert the aluminum strips into an Ehrlenmeyer flask and plug the flask as shown in the diagram. Be careful not to damage the aluminum strips. (6) Inflate a balloon. (7) Rub the balloon against a section of wool or cotton clothing. (8) Place the balloon near the point of the nail. (9) Record your observations. (10) Touch the point of the nail with a piece of paper or glass rod. (11) Record your observations.

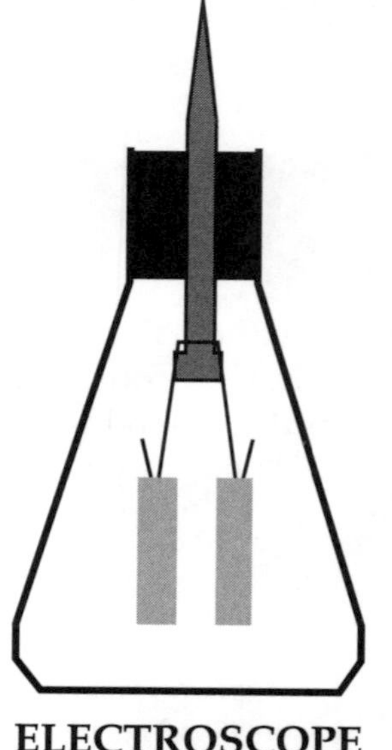

ELECTROSCOPE

PS15 **Lesson #2**

STATIC AND CURRENT ELECTRICITY

Work Date: ____/____/____

LESSON OBJECTIVE

Students will draw electrical circuit diagrams for a simple series circuit and explain the relationship between amperage, voltage, and resistance in a circuit.

Classroom Activities

On Your Mark!

Draw a battery, a light bulb, a switch, and wires as shown in Illustration A. Identify the parts for students and have them make a small copy of the illustration on Journal Sheet #2. Inform them that electrical engineers do not plan circuits in the same way an artist prepares a "real life" drawing. The electrical engineer uses a circuit diagram. Have them copy the simple circuit diagram shown in Illustration B. Identify the symbols and tell them to use the same symbols in completing their Homework.

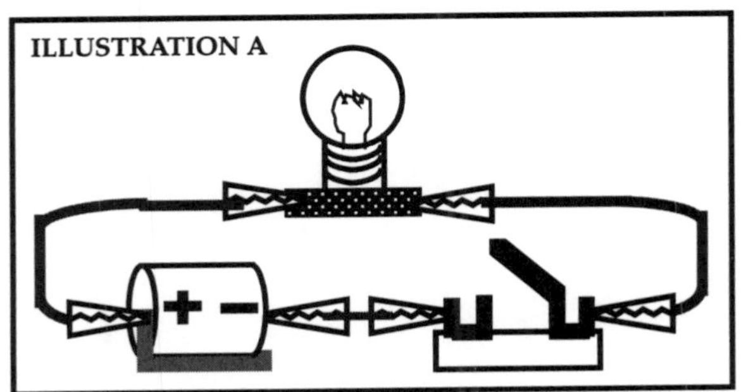

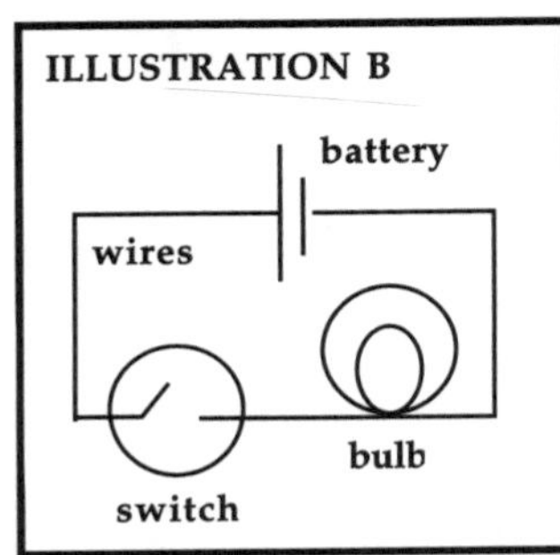

Get Set!

Have students record the following definitions on Journal Sheet #2:

coulomb: unit of electrical *charge* equal to the charge on 6×10^{23} electrons
ampere: unit of electrical *current* equal to the flow of one coulomb of charge past a given point in one second
volt: unit of electrical *pressure* equal to the amout of energy required to push one coulomb of charge through a resistor (i.e., a joule per coulomb)
ohm: unit of electrical *resistance* equal to the amount of pressure required to sustain one ampere of current in an electrical conductor (i.e., a volt per ampere)

Introduce students to **Ohm's Law: $I = V \div R$**. Explain the meaning of each variable and work through several examples on the board to make sure students understand how to use the formula. Explain that there is a **direct relationship** between amperage and voltage in an electrical circuit and an **inverse relationship** between amperage and resistance.

Go!

Assist students in constructing the circuit diagrams in Figure B on Journal Sheet #2. They will find that more batteries increase the brightness of the bulbs (i.e, direct relationship between amperage and voltage) while more bulbs decreases their brightness (i.e., inverse relationship between amperage and resistance). Have them unscrew one of the bulbs from the circuit. They will discover that the other bulb goes out. In a series circuit the following is true: "If one appliance blows, then they all blow!"

Materials

D-cell batteries, battery holders and alligator clips (or tape), insulated thin-guage copper wire, switches, flashlight bulbs of equal resistances and sockets to fit

Name: ____________________ Period:________ Date: ___/___/___

PS15 Journal Sheet #2

STATIC AND CURRENT ELECTRICITY

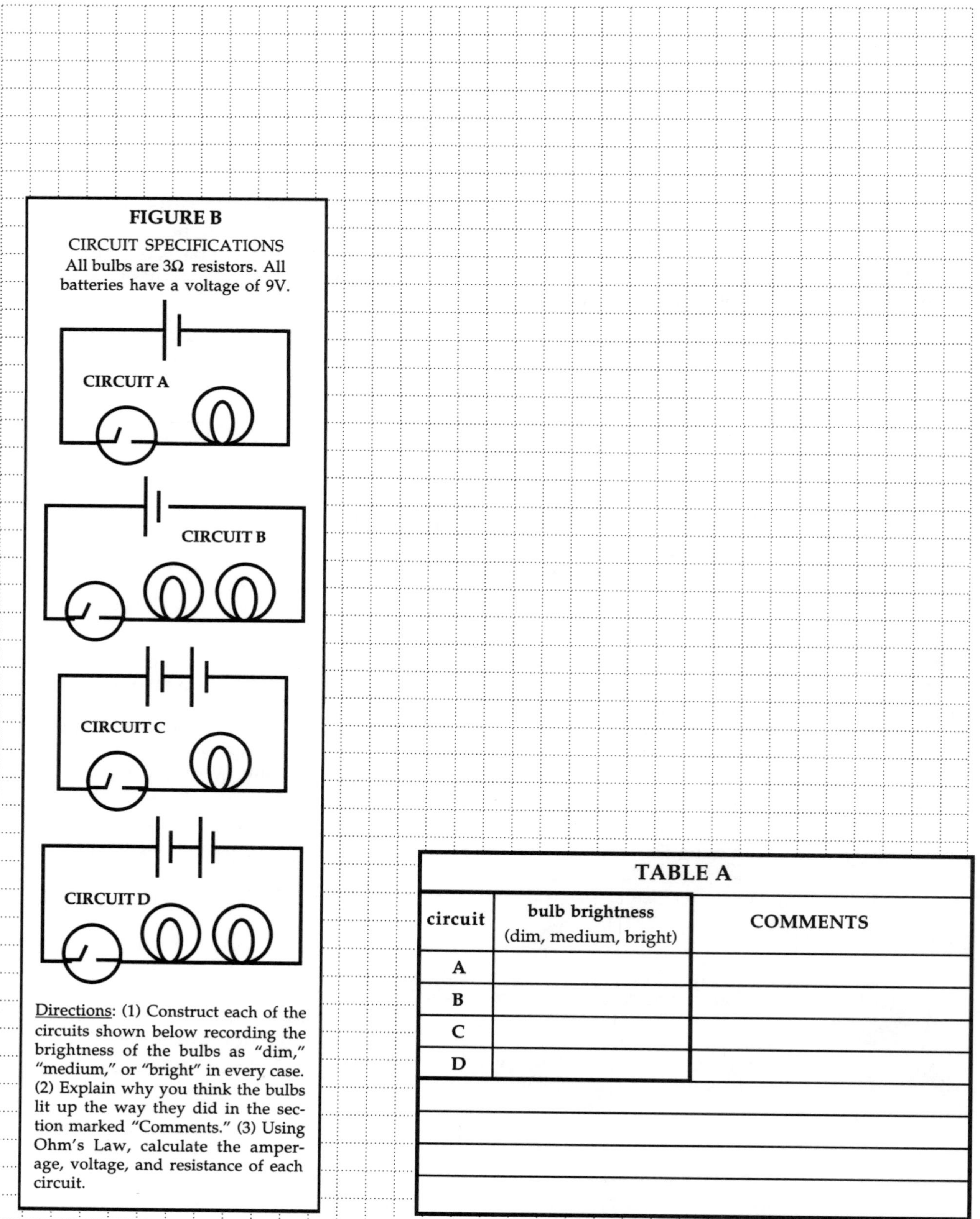

Directions: (1) Construct each of the circuits shown below recording the brightness of the bulbs as "dim," "medium," or "bright" in every case. (2) Explain why you think the bulbs lit up the way they did in the section marked "Comments." (3) Using Ohm's Law, calculate the amperage, voltage, and resistance of each circuit.

TABLE A

circuit	bulb brightness (dim, medium, bright)	COMMENTS
A		
B		
C		
D		

STATIC AND CURRENT ELECTRICITY

Work Date: ____/____/____

LESSON OBJECTIVE

Students will construct a simple parallel circuit and discuss the advantages of this kind of circuit over a series circuit.

Classroom Activities

On Your Mark!

Review the discoveries made in Lesson #2. In a series circuit, there is only one pathway for the current to take. If one of the appliances in a series circuit blows out, then all of the appliances blow out. That is because there is no way for the current to complete its "circular journey" from the negative terminal of the battery to the positive terminal. This principle is the origin of the term "circuit."

Get Set!

Propose the following idea: "What would happen if we gave the current an alternative pathway through which to flow, like a freeway off ramp gives drivers an alternative path to take if the freeway gets too crowded?"

Go!

Assist students in constructing the circuit diagrams in Figure C on Journal Sheet #3. They will find that more batteries increases the brightness of the bulbs as it did in their series circuit constructed in Lesson #2. However, an increase in the number of bulbs does not cause each bulb to grow dim. In addition, unscrewing one of the bulbs in the parallel circuit does not affect the brightness of the other bulbs. Ask them to consider how the appliances in their home are hooked up. Are they connected in series or parallel? Answer: Parallel. If a light bulb in their room blows, their CD player still remains on! Help them to calculate the total resistance of their parallel circuit as explained in the answer to Homework Problem #2 on the Teacher's Agenda and Content Notes.

Materials

D-cell batteries, battery holders and alligator clips (or tape), insulated thin-gauge copper wire, switches, flashlight bulbs of equal resistances and sockets to fit

Name: ______________________ Period: ________ Date: ___/___/___

PS15 Journal Sheet #3

STATIC AND CURRENT ELECTRICITY

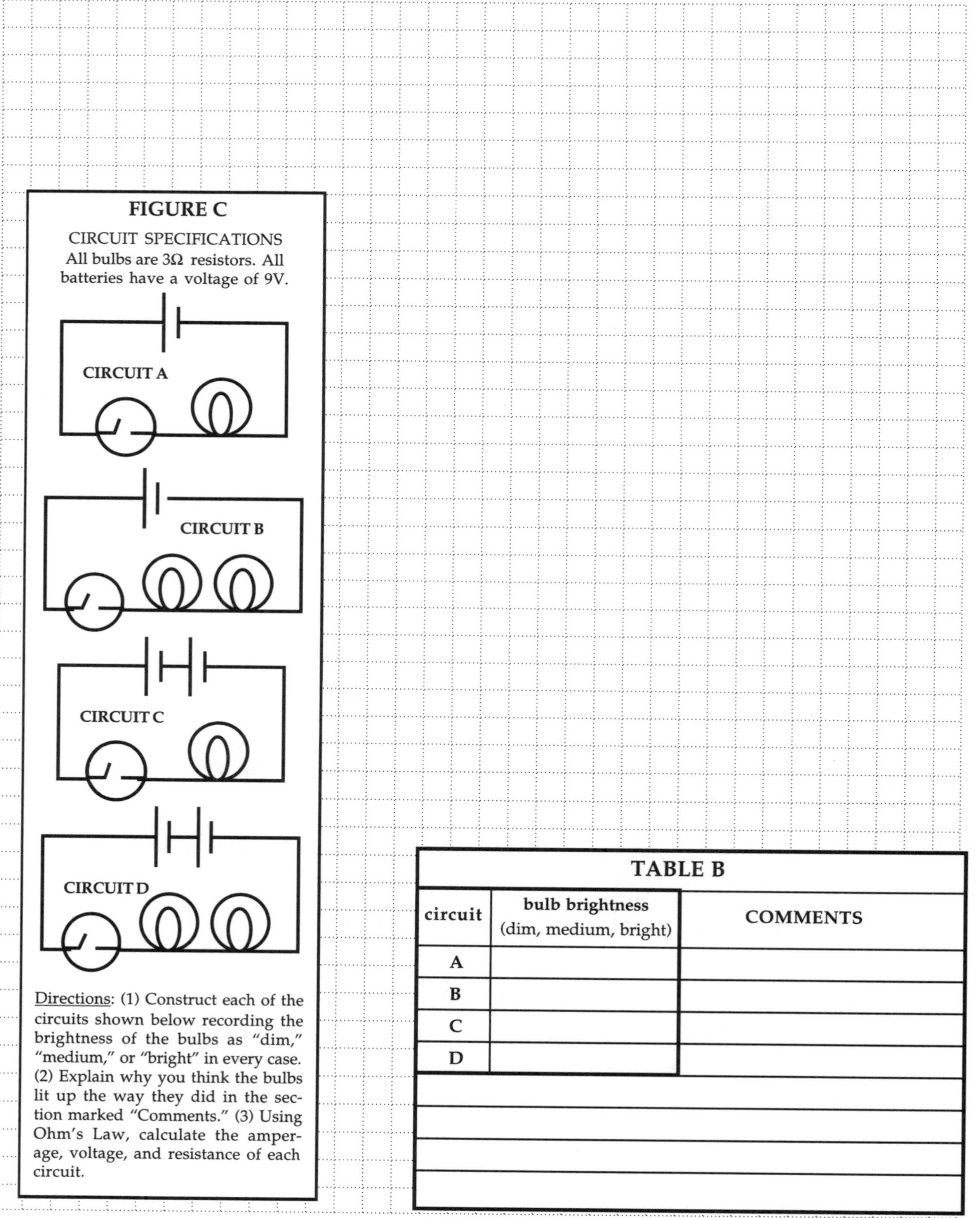

FIGURE C

CIRCUIT SPECIFICATIONS
All bulbs are 3Ω resistors. All batteries have a voltage of 9V.

Directions: (1) Construct each of the circuits shown below recording the brightness of the bulbs as "dim," "medium," or "bright" in every case. (2) Explain why you think the bulbs lit up the way they did in the section marked "Comments." (3) Using Ohm's Law, calculate the amperage, voltage, and resistance of each circuit.

TABLE B

circuit	bulb brightness (dim, medium, bright)	COMMENTS
A		
B		
C		
D		

STATIC AND CURRENT ELECTRICITY

Work Date: ____/____/____

LESSON OBJECTIVE

Students will construct a simple electric motor.

Classroom Activities

On Your Mark!

Discuss the difference between a **motor** and a **generator** as it appears in the Fact Sheet. Lead a discussion of the discoveries and contributions made by inventors **Michael Faraday** (b. 1791; d. 1867) and **Thomas Edison** (b. 1847; d. 1931).

Get Set!

Prior to the start of class, build a working model of the simple electric motor shown in Figure D on Journal Sheet #4 according the the following directions:

1. Tape two small iron nails together point to point.
2. Make your electromagnet. Starting at the middle of the nails begin winding tight coils of thin-gauge insulated copper wire around the nails toward the head of one nail. When you reach the head of the first nail return to the head of the other nail by making a second layer of coiling over the first layer. When you reach the head of the second nail continue winding back toward the center. *Be sure to continue coiling in the same direction. Do not reverse direction or criss-cross coils in a sloppy fashion.*
3. Strip the ends of the wire and use masking tape to cover the top half of one exposed wire as shown in Figure D. The bottom half of the wire must remain exposed in order for current to flow through the electromagnet touching the paper clips.
4. Bend two paper clips as shown in Figure D and insert them into a piece of cardboard as shown.
5. Insert the wire ends of the electromagnet into the loops on the paper clip and balance the electromagnet so that it spins freely.
6. Connect the battery (or batteries) and switch as shown and place either pole of a bar magnet close to the end of the electromagnet without touching it.

When current flows through the electromagnet it becomes polarized and, with a push to help it get started, will be repelled by the stationary magnet and start to spin. As it rotates, current will be turned on and off by the tape placed over the half-taped wire. This will prevent the electromagnet from ever nearing the stationary magnet with an opposite (i.e., attracting) pole.

Go!

Explain the entire procedure for constructing the motor to students. Impress them with the notion that attention to detail and craftsmanship are the keys to success in this project.

Materials

D-cell batteries, battery holders and alligator clips (or tape), insulated thin-gauge copper wire, switches, iron nails, masking tape, paper clips, cardboard

Name: ______________________ Period: ________ Date: ___/___/___

PS15 Journal Sheet #4

STATIC AND CURRENT ELECTRICITY

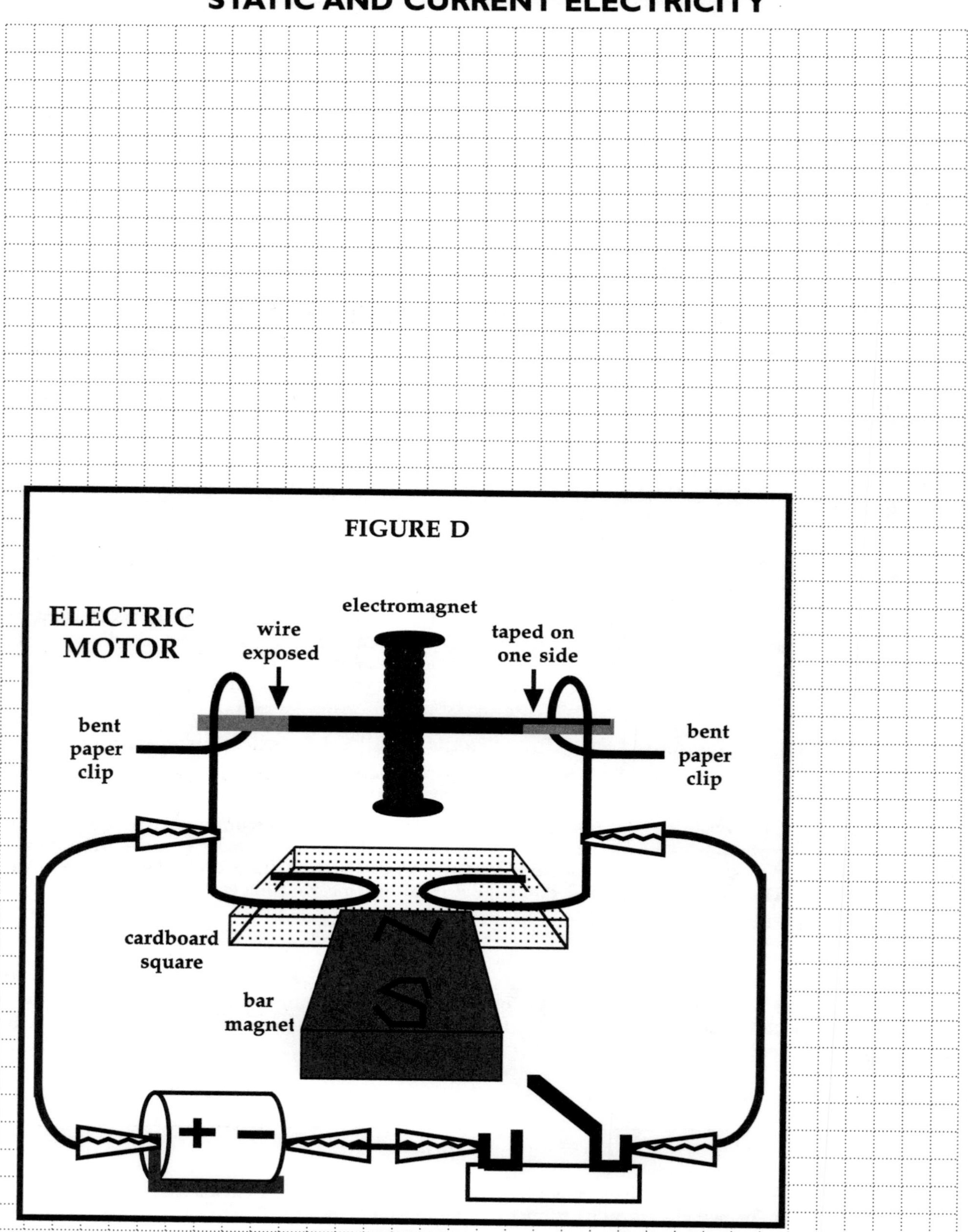

Name: ______________________ Period:________ Date: ____/____/____

PS15 Review Quiz

Directions: Keep your eyes on your own work.
Read all directions and questions carefully.
THINK BEFORE YOU ANSWER!
Watch your spelling, be neat, and do the best you can.

CLASSWORK	(~40):	_______
HOMEWORK	(~20):	_______
CURRENT EVENT	(~10):	_______
TEST	(~30):	_______
TOTAL	(~100):	_______

(A ≥ 90, B ≥ 80, C ≥ 70, D ≥ 60, F < 60)

LETTER GRADE: _______

TEACHER'S COMMENTS: ______________________

STATIC AND CURRENT ELECTRICITY

TRUE–FALSE FILL-IN: If the statement is true, write the word TRUE. If the statement is false, change the underlined word to make the statement true. *10 points*

______________ 1. Count Alessandro Dracula invented the first battery.

______________ 2. Electric charges are measured in amperes.

______________ 3. The flow of electric charges is measured in volts.

______________ 4. The amount of force required to "push" electric charges through a wire is called resistance.

______________ 5. An appliance's resistance to the flow of electricity is measured in ohms.

______________ 6. The relationship between amperage, voltage, and resistance can be summarized by Newton's Law.

______________ 7. Electricity flowing through a wire produces a magnetic field.

______________ 8. The core of the earth, which is made of salt, behaves like a magnet.

______________ 9. A generator transforms electrical energy to kinetic energy.

______________ 10. A motor transforms kinetic energy to electrical energy.

List the names of five scientists who added to our understanding of how electricity works. In a phrase, describe that scientist's major invention or contribution.

NAME	INVENTION OR CONTRIBUTION
______________	______________________________
______________	______________________________
______________	______________________________
______________	______________________________
______________	______________________________

PROBLEM

Directions: Use Figure I to answer questions #11 through #15.

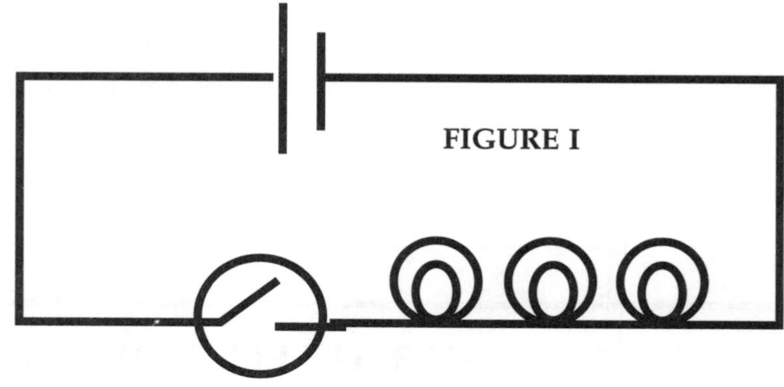

CIRCUIT DIAGRAM SPECIFICATIONS

Battery: 12 volts
Bulbs: 2 ohms resistance each
Switch: resistance negligible

11. Is this circuit a series or parallel circuit?

____

12. If one bulb in this electric circuit blew out, what would happen to the other bulbs?

____

13. What is the total resistance of this circuit?

____

14. If another 12-volt battery were put in series with the first battery in the circuit, would the bulbs be brighter or dimmer?

____

15. How much current is flowing through the circuit in Figure I? (*Note*: Show formulas and all math calculations!)

____

Final Answer: _____________

_______________________ Student's Signature

_______________________ Parent's Signature

____/____/____ Date

APPENDIX

Name: ______________________________ **Period:**_______ **Date:** ___/___/___

EXTRA JOURNAL SHEET

FACT SHEET TITLE: ___________________________________ **Lesson #** __________

NAME: ______________________________ **PERIOD:** ______ **DATE:** ___ / ___ / ___

Keep this Grade Roster in the Science Section of your notebook

Date	Journal Points	Homework Points	Current Events Points	Quiz Points	Total Points	Letter Grade	Initials

How to Calculate Your Grade Point Average

Your Report Card grades in this class will be awarded to you according to your grade point average or GPA. You can calculate your GPA whenever you like to find out exactly how you are doing in this class.

First, award each of your weekly grades the following credits: each A is worth 4 credits; each B is worth 3 credits; each C is worth 2 credits; each D is worth 1 credit; and each F is worth 0.

Add your total credits earned. Then, divide by the number of packets listed on your Grade Roster and round the decimal result to the nearest tenths place. Your overall Letter Grade is assigned according to the following GPA values:

A+ ≥ 4.0	A ≥ 3.7	A– ≥ 3.4
B+ ≥ 3.1	B ≥ 2.8	B– ≥ 2.5
C+ ≥ 2.2	C ≥ 1.9	C– ≥ 1.6
D+ ≥ 1.3	D ≥ 1.0	D– ≥ 0.7
	F < 0.7	

FOR EXAMPLE:

John has completed five weeks of school and entered his grades from five packets on his Grade Roster. His grades are as follows: first week, A; second week, B; third week, C; fourth week, C; and fifth week, D.

John awards himself the correct amount of credit for each of his grades.

A	earns	4 credits
B	earns	3 credits
C	earns	2 credits
C	earns	2 credits
D	earns	1 credits
Total	earned is	12 credits

John divides his total credits earned by 5 (the number of packets on his Grade Roster).

12 divided by 5 equals 2.4

John's grade point average, or GPA, is 2.4. Referring to the grades shown above, John knows that he has a C+ in Science thusfar, because 2.4 is greater than 2.2 (C+) but less than 2.5 (B–).

Using Current Events to Integrate Science Instruction Across Content Areas

Science does not take place in a vacuum. Scientists, like other professionals, are influenced by the economic and political realities of their time. In addition, the ideological and technological advances made by science can influence the economic and political structure of society—for better or worse. It is therefore essential that students have an awareness of the day-to-day science being done at laboratories around the world, and the important work being reported by an international news media.

Most State Departments of Education make **Current Events** a regular part of their state science frameworks. The science instructor can use newspaper, magazine, and television reports to keep their students informed about the advances and controversies stemming from research in the many scientific disciplines. Teachers can also use current events to integrate science instruction across the curriculum.

Set aside a class period to show students how to prepare a **science** or **technology current event.** They can do this on a single sheet of standard looseleaf paper. You may require pupils to read all or part of a science/technology article, depending on their reading level. Have them practice summarizing the lead and one or more paragraphs of the article in their own words. Advise them to keep a **thesaurus** on hand or to use the dictionary/thesaurus stored in their personal computer at home. Tell students to find synonyms they can use to replace most of the vocabulary words used by the author of the article. This activity will help them to expand their vocabulary and improve their grammar skills. Show students how to properly trim and paste the article's title and first few paragraphs on the front of a standard piece of looseleaf. They should write their summary on the opposite side of the page so that the article is visible to their classmates when they present their findings orally to the class. Allow students to make a report that summarizes a newsworthy item they may have heard about on television. The latter report should be accompanied by the signature of a parent/guardian to insure the accuracy of the information being presented.

Students' skills at public speaking are sure to improve if they are given an opportunity to share their current event with the rest of the class.. Current events can be shared after the end-of-the-unit Review Quiz or whenever the clock permits at the end of a lesson that has been completed in a timely fashion. Select students at random to make their presentations by drawing lots, or ask for volunteers who might be especially excited about their article. Take time to discuss the ramifications of the article and avoid the temptation to express your personal views or bias. Remain objective and give students the opportunity to express their views and opinions. Encourage them to base their views on fact—not superstition or prejudice. Should the presentation turn into a debate, set aside a few minutes later in the week, giving students time to prepare what they would like to say. Model courtesy and respect for all points of view and emphasize the proper use of the English language in all modes of presentation—both written and oral.

BIO-DATA CARDS

INSTRUCTIONS TO TEACHERS
Xerox and cut out the Bio-Data Cards below and keep them in a handy file. Instruct students to choose one card and neatly glue it to the front of a 5″ × 8″ index card. They can use the school or public library to find out more about the scientist they have chosen. On the back of the index card they can draw a cartoon, write a poem or short paragraph that illustrates an important event in the life of this famous personality.

BIO-DATA CARD

ANDRÉ MARIE AMPERE
(born, 1775; died, 1836)
nationality
French
contribution to science
discovered the rule for describing the direction of a magneic field that surrounds an electric current

BIO-DATA CARD

ARCHIMEDES
(born, 287 B.C.; died, 212 B.C.
nationality
Greek
contribution to science
discovered why objects float in a fluid making it possible to build better ships and float away in a hot-air balloon

BIO-DATA CARD

KARL BENZ
(born, 1844; died, 1929)
nationality
German
contribution to science
developed the first automobile driven by an internal combustion engine

BIO-DATA CARD

DANIEL BERNOULLI
(born, 1700; died, 1782)
nationality
Swiss
contribution to science
discovered that fluids moving fast across a surface exert less pressure on that surface, making human flight possible

BIO-DATA CARD

HENRY CAVENDISH
(born, 1731; died, 1810)
nationality
English
contribution to science
the first to measure the mass and density of the earth and determine the composition of water

BIO-DATA CARD

NIKOLAUS COPERNICUS
(born, 1473; died, 1543)
nationality
Polish
contribution to science
showed that the sun—not the earth—was the center of the solar system

BIO-DATA CARD

CHARLES COULOMB
(born, 1736; died, 1806)
nationality
French
contribution to science
invented a "torsion balance" that allowed him to measure the force of attraction between charged objects

BIO-DATA CARD

ALBERT EINSTEIN
(born, 1879; died, 1955)
nationality
German–American
contribution to science
showed that the motion of objects was relative and that light waves bend in a strong gravitational field

INSTRUCTIONS TO TEACHERS
Xerox and cut out the Bio-Data Cards below and keep them in a handy file. Instruct students to choose one card and neatly glue it to the front of a 5″ × 8″ index card. They can use the school or public library to find out more about the scientist they have chosen. On the back of the index card they can draw a cartoon, write a poem or short paragraph that illustrates an important event in the life of this famous personality.

BIO-DATA CARD

MICHAEL FARADAY
(born, 1791; died, 1867)
nationality
English
contribution to science
invented the electric motor and transformer

BIO-DATA CARD

HENRY FORD
(born, 1863; died, 1947)
nationality
American
contribution to science
developed the assembly line to produce automobiles by mass-production, making them less expensive to buy

BIO-DATA CARD

BENJAMIN FRANKLIN
(born, 1706; died, 1790)
nationality
American
contribution to science
invented the lightning rod

BIO-DATA CARD

GALILEO GALILEI
(born, 1564; died, 1642)
nationality
Italian
contribution to science
demonstrated that all falling objects fall to earth at the same rate of speed

BIO-DATA CARD

WILLIAM GILBERT
(born, 1540; died, 1603)
nationality
English
contribution to science
explained how the earth's magnetic field behaves like a bar magnet stuck between the North and South Poles

BIO-DATA CARD

HERO OF ALEXANDRIA
(lived in the first century A.D.)
nationality
Greek–Egyptian
contribution to science
made gears able to lift many times their own weight and built a steam engine to show the power of moving gases

BIO-DATA CARD

HEINRICH RUDOLF HERTZ
(born, 1857; died, 1894)
nationality
German
contribution to science
showed that electric waves could move through air which led to the discovery of radio waves

BIO-DATA CARD

CHRISTIAAN HUYGENS
(born, 1629; died, 1695)
nationality
Dutch
contribution to science
invented the first pendulum clock

INSTRUCTIONS TO TEACHERS
Xerox and cut out the Bio-Data Cards below and keep them in a handy file. Instruct students to choose one card and neatly glue it to the front of a 5″ × 8″ index card. They can use the school or public library to find out more about the scientist they have chosen. On the back of the index card they can draw a cartoon, write a poem or short paragraph that illustrates an important event in the life of this famous personality.

BIO-DATA CARD

JAMES PRESCOTT JOULE
(born, 1818; died, 1889)
nationality
English
contribution to science
discovered that energy cannot be created nor destroyed but is only transformed from one form to another

BIO-DATA CARD

WILLIAM THOMSON KELVIN
(born, 1824; died, 1907)
nationality
Irish
contribution to science
introduced the kelvin scale to measure the coldest objects in the universe

BIO-DATA CARD

GEORGES LECLANCHÉ
(born, 1839; died, 1882)
nationality
French
contribution to science
made the first dry cell battery using starch to paste the other chemicals together

BIO-DATA CARD

GOTTFRIED LEIBNIZ
(born, 1646; died, 1716)
nationality
German
contribution to science
devised the first calculating machine able to multiply and divide

BIO-DATA CARD

JEAN JOSEPH LENOIR
(born, 1822; died, 1900)
nationality
Belgian
contribution to science
designed and built the first gas-powered internal combustion engine

BIO-DATA CARD

HANS LIPPERSHEY
(born, 1570; died, 1619)
nationality
Dutch
contribution to science
patented the first refracting telescope which he called a "looker"

BIO-DATA CARD

KIRKPATRICK MACMILLAN
(born, 1813; died, 1878)
nationality
Scottish
contribution to science
put a "hobby-horse" on wheels, cranks, and peddles and invented the first practical bicycle

BIO-DATA CARD

JAMES CLERK MAXWELL
(born, 1831; died, 1879)
nationality
Scottish
contribution to science
developed electromagnetic field theory making possible the invention of the telegraph and telephone

INSTRUCTIONS TO TEACHERS
Xerox and cut out the Bio-Data Cards below and keep them in a handy file. Instruct students to choose one card and neatly glue it to the front of a 5″ × 8″ index card. They can use the school or public library to find out more about the scientist they have chosen. On the back of the index card they can draw a cartoon, write a poem or short paragraph that illustrates an important event in the life of this famous personality.

BIO-DATA CARD

ANDRÉ MICHELIN

(born, 1853; died, 1931)

nationality
French

contribution to science
invented the first "air-filled" pneumatic tire for use on bicycles and automobiles

BIO-DATA CARD

ALBERT A. MICHELSON

(born, 1852; died, 1931)

nationality
German–American

contribution to science
invented an interferometer and proved that light waves travel through the vacuum of space at a constant speed

BIO-DATA CARD

EDWARD W. MORLEY

(born, 1838; died, 1923)

nationality
American

contribution to science
made the first accurate measure of the densities of oxygen and hydrogen and worked with Albert A. Michelson

BIO-DATA CARD

SIR ISAAC NEWTON

(born, 1642; died, 1727)

nationality
English

contribution to science
made space flight possible with his discovery of three Laws of Motion and the Universal Law of Gravity

BIO-DATA CARD

ALFRED NOBEL

(born, 1833; died, 1896)

nationality
Swedish

contribution to science
invented dynamite and founded the Nobel Prize to promote peace and good will throughout the world

BIO-DATA CARD

HANS CHRISTIAN OERSTED

(born, 1777; died, 1851)

nationality
Danish

contribution to science
discovered that electric charges in motion produce a magnetic field

BIO-DATA CARD

GEORG SIMON OHM

(born, 1789; died, 1854)

nationality
German

contribution to science
showed the relationship between amperage, voltage, and resistance in an electrical circuit

BIO-DATA CARD

NIKOLAUS OTTO

(born, 1832; died, 1891)

nationality
German

contribution to science
built the first modern 4-stroke internal combustion engine

INSTRUCTIONS TO TEACHERS
Xerox and cut out the Bio-Data Cards below and keep them in a handy file. Instruct students to choose one card and neatly glue it to the front of a 5″ × 8″ index card. They can use the school or public library to find out more about the scientist they have chosen. On the back of the index card they can draw a cartoon, write a poem or short paragraph that illustrates an important event in the life of this famous personality.

BIO-DATA CARD

THOMAS SAVERY
(born, 1650; died, 1715)

nationality
English

contribution to science
invented the first steam-driven pump, boiling water over an open flame

BIO-DATA CARD

JAMES STARLEY
(born, 1830; died, 1881)

nationality
English

contribution to science
considered the "father of the bicycle industry" for improving bicycle design with gears, chains, and spoke-wheels

BIO-DATA CARD

WILLIAM STURGEON
(born, 1783; died, 1850)

nationality
English

contribution to science
made the first electromagnet and invented the galvanometer

BIO-DATA CARD

EVANGELISTA TORRICELLI
(born, 1608; died, 1647)

nationality
Italian

contribution to science
invented the first mercury barometer to measure atmospheric pressure

BIO-DATA CARD

ROBERT van de GRAAFF
(born, 1901; died, 1967)

nationality
American

contribution to science
invented a powerful generator that could be used to study the movement of static electric charges

BIO-DATA CARD

ALESSANDRO VOLTA
(born, 1745; died, 1827)

nationality
Italian

contribution to science
invented the first battery by separating silver and zinc metal plates with pieces of cloth soaked in acid

BIO-DATA CARD

JAMES WATT
(born, 1736; died, 1819)

nationality
Scottish

contribution to science
developed an improved design for a steam engine that could be put to commercial use

BIO-DATA CARD

WILHEM EDUARD WEBER
(born, 1804; died, 1891)

nationality
German

contribution to science
developed sensitive magnetometers to measure magnetic fields and electrical currents